CW01151612

Offshore Compliant Platforms

Wiley-ASME Press Series

Corrosion and Materials in Hydrocarbon Production: A Compendium of Operational and Engineering Aspects
Bijan Kermani, Don Harrop

Design and Analysis of Centrifugal Compressors
Rene Van den Braembussche

Case Studies in Fluid Mechanics with Sensitivities to Governing Variables
M. Kemal Atesmen

The Monte Carlo Ray-Trace Method in Radiation Heat Transfer and Applied Optics
J. Robert Mahan

Dynamics of Particles and Rigid Bodies: A Self-Learning Approach
Mohammed F. Daqaq

Primer on Engineering Standards, Expanded Textbook Edition
Maan H. Jawad, Owen R. Greulich

Engineering Optimization: Applications, Methods, and Analysis
R. Russell Rhinehart

Compact Heat Exchangers: Analysis, Design and Optimization Using FEM and CFD Approach
C. Ranganayakulu, Kankanhalli N. Seetharamu

Robust Adaptive Control for Fractional-Order Systems with Disturbance and Saturation
Mou Chen, Shuyi Shao, Peng Shi

Robot Manipulator Redundancy Resolution
Yunong Zhang, Long Jin

Stress in ASME Pressure Vessels, Boilers, and Nuclear Components
Maan H. Jawad

Combined Cooling, Heating, and Power Systems: Modeling, Optimization, and Operation
Yang Shi, Mingxi Liu, Fang Fang

Applications of Mathematical Heat Transfer and Fluid Flow Models in Engineering and Medicine
Abram S. Dorfman

Bioprocessing Piping and Equipment Design: A Companion Guide for the ASME BPE Standard
William M. (Bill) Huitt

Nonlinear Regression Modeling for Engineering Applications: Modeling, Model Validation, and Enabling Design of Experiments
R. Russell Rhinehart

Geothermal Heat Pump and Heat Engine Systems: Theory and Practice
Andrew D. Chiasson

Fundamentals of Mechanical Vibrations
Liang-Wu Cai

Introduction to Dynamics and Control in Mechanical Engineering Systems
Cho W.S. To

Offshore Compliant Platforms

Analysis, Design, and Experimental Studies

Srinivasan Chandrasekaran
Department of Ocean Engineering
Indian Institute of Technology Madras
Tamil Nadu
India

R. Nagavinothini
Department of Structures for Engineering and Architecture
University of Naples Federico II
Naples
Italy

This Work is a co-publication between John Wiley & Sons Ltd and ASME Press

WILEY

ASME PRESS

This edition first published 2020
© 2020 John Wiley & Sons Ltd

This Work is a co-publication between John Wiley & Sons Ltd and ASME Press

All rights reserved. No part of this publication may be reproduced, stored in a retrieval system, or transmitted, in any form or by any means, electronic, mechanical, photocopying, recording or otherwise, except as permitted by law. Advice on how to obtain permission to reuse material from this title is available at http://www.wiley.com/go/permissions.

The right of Srinivasan Chandrasekaran and R. Nagavinothini to be identified as the authors of this work has been asserted in accordance with law.

Registered Offices
John Wiley & Sons, Inc., 111 River Street, Hoboken, NJ 07030, USA
John Wiley & Sons Ltd, The Atrium, Southern Gate, Chichester, West Sussex, PO19 8SQ, UK

Editorial Office
The Atrium, Southern Gate, Chichester, West Sussex, PO19 8SQ, UK

For details of our global editorial offices, customer services, and more information about Wiley products visit us at www.wiley.com.

Wiley also publishes its books in a variety of electronic formats and by print-on-demand. Some content that appears in standard print versions of this book may not be available in other formats.

Limit of Liability/Disclaimer of Warranty
MATLAB® is a trademark of The MathWorks, Inc. and is used with permission. The MathWorks does not warrant the accuracy of the text or exercises in this book. This work's use or discussion of MATLAB® software or related products does not constitute endorsement or sponsorship by The MathWorks of a particular pedagogical approach or particular use of the MATLAB® software.

In view of ongoing research, equipment modifications, changes in governmental regulations, and the constant flow of information relating to the use of experimental reagents, equipment, and devices, the reader is urged to review and evaluate the information provided in the package insert or instructions for each chemical, piece of equipment, reagent, or device for, among other things, any changes in the instructions or indication of usage and for added warnings and precautions. While the publisher and authors have used their best efforts in preparing this work, they make no representations or warranties with respect to the accuracy or completeness of the contents of this work and specifically disclaim all warranties, including without limitation any implied warranties of merchantability or fitness for a particular purpose. No warranty may be created or extended by sales representatives, written sales materials or promotional statements for this work. The fact that an organization, website, or product is referred to in this work as a citation and/or potential source of further information does not mean that the publisher and authors endorse the information or services the organization, website, or product may provide or recommendations it may make. This work is sold with the understanding that the publisher is not engaged in rendering professional services. The advice and strategies contained herein may not be suitable for your situation. You should consult with a specialist where appropriate. Further, readers should be aware that websites listed in this work may have changed or disappeared between when this work was written and when it is read. Neither the publisher nor authors shall be liable for any loss of profit or any other commercial damages, including but not limited to special, incidental, consequential, or other damages.

Library of Congress Cataloging-in-Publication Data

Names: Chandrasekaran, Srinivasan, author. | Nagavinothini, R., 1991– author.
Title: Offshore compliant platforms : analysis, design, and experimental studies / Srinivasan Chandrasekaran, Department of Ocean Engineering, Indian Institute of Technology Madras, Tamil Nadu, India, R. Nagavinothini, Department of Structures for Engineering and Architecture, University of Naples Federico II, Naples, Italy.
Description: First edition. | Hoboken, NJ : John Wiley & Sons, Inc., 2020. | Series: Wiley-ASME Press Series | Includes bibliographical references and index.
Identifiers: LCCN 2019050441 (print) | LCCN 2019050442 (ebook) | ISBN 9781119669777 (hardback) | ISBN 9781119669784 (adobe pdf) | ISBN 9781119669807 (epub)
Subjects: LCSH: Compliant platforms.
Classification: LCC TC1700 .C44 2020 (print) | LCC TC1700 (ebook) | DDC 627/.98–dc23
LC record available at https://lccn.loc.gov/2019050441
LC ebook record available at https://lccn.loc.gov/2019050442

Cover Design: Wiley
Cover Image: © paranyu pithayarungsarit/Getty Images

Set in 9.5/12.5pt STIXTwoText by SPi Global, Pondicherry, India

Printed and bound by CPI Group (UK) Ltd, Croydon, CR0 4YY

10 9 8 7 6 5 4 3 2 1

Contents

List of Figures *ix*
List of Tables *xiii*
Foreword by Professor Purnendu K. Das *xv*
Foreword by Dr. Atmanand N.D. *xvii*
Series Preface *xix*
Preface *xxi*

1 **Common Compliant Platforms** *1*
1.1 Introduction *1*
1.2 Tension Leg Platforms *8*
1.3 Guyed Tower and Articulated Tower *19*
1.4 Floating Structures *21*
1.5 Response Control Strategies *24*
1.5.1 Active Control Algorithm *25*
1.5.2 Semi-Active Control Algorithm *25*
1.5.3 Passive Control Algorithm *26*
1.5.4 Friction Dampers *27*
1.5.5 Metallic Yield Dampers *27*
1.5.6 Viscous Fluid Dampers *27*
1.5.7 Tuned Liquid Dampers *29*
1.5.8 Tuned Liquid Column Damper *30*
1.6 Tuned Mass Dampers *31*
1.7 Response Control of Offshore Structures *36*
1.8 Response Control of TLPs Using TMDs: Experimental Investigations *38*
1.9 Articulated Towers *44*
1.10 Response Control of ATs: Analytical Studies *48*
1.11 Response Control of ATs: Experimental Studies *52*
1.11.1 MLAT Without a TMD *53*
1.11.2 MLAT with a TMD *56*

2	**Buoyant Leg Storage and Regasification Platforms**	*59*
2.1	Background Literature *60*	
2.1.1	Buoyant Leg Structures *62*	
2.1.2	Floating Production and Processing Platforms *63*	
2.2	Experimental Setup *64*	
2.3	Experimental Investigations *65*	
2.4	Numerical Studies *72*	
2.5	Critical Observations *76*	
2.6	Stability Analysis of the BLSRP *85*	
2.7	Fatigue Analysis of the BLSRP *90*	

3	**New-Generation Platforms: Offshore Triceratops**	*95*
3.1	Introduction *95*	
3.2	Environmental Loads *96*	
3.2.1	Regular Waves *96*	
3.2.2	Random Waves *97*	
3.2.3	Wind *98*	
3.2.4	Currents *100*	
3.3	Fatigue Analysis of Tethers *101*	
3.4	Response to Regular Waves *104*	
3.5	Response to Random Waves *108*	
3.6	Response to Combined Actions of Wind, Waves, and Current *113*	
3.6.1	Deck Response *116*	
3.6.2	Buoyant Leg Response *120*	
3.6.3	Tether Tension Variation *122*	
3.7	Summary *123*	

4	**Triceratops Under Special Loads** *125*	
4.1	Introduction *125*	
4.1.1	Ice Load *126*	
4.1.2	Impact Load Due to Ship Platform Collisions *129*	
4.1.3	Hydrocarbon Fires *131*	
4.2	Continuous Ice Crushing *134*	
4.2.1	The Korzhavin Equation *135*	
4.2.2	Continuous Ice Crushing Spectrum *136*	
4.3	Response to Continuous Ice Crushing *138*	
4.3.1	Response to Ice Loads *139*	
4.3.1.1	Deck and Buoyant Leg Responses *139*	
4.3.1.2	Tether Response *140*	

4.3.2	Effect of Ice Parameters	*140*
4.3.2.1	Ice Thickness	*140*
4.3.2.2	Ice Crushing Strength	*143*
4.3.2.3	Ice Velocity	*144*
4.3.3	Comparison of Ice- and Wave-Induced Responses	*145*
4.4	Response to Impact Loads	*147*
4.4.1	Parametric Studies	*151*
4.4.1.1	Indenter Size	*151*
4.4.1.2	Collision Zone Location	*152*
4.4.1.3	Indenter Shape	*153*
4.4.1.4	Number of Stringers	*154*
4.4.2	Impact Response in the Arctic Region	*154*
4.5	Deck Response to Hydrocarbon Fires	*156*
4.6	Summary	*158*
5	**Offshore Triceratops: Recent Advanced Applications**	*161*
5.1	Introduction	*161*
5.2	Wind Turbines	*161*
5.3	Wind Power	*163*
5.4	Evolution of Wind Turbines	*163*
5.5	Conceptual Development of the Triceratops-Based Wind Turbine	*164*
5.6	Support Systems for Wind Turbines	*164*
5.6.1	Spar Type	*165*
5.6.2	TLP Type	*165*
5.6.3	Pontoon (Barge) Type	*165*
5.6.4	Semi-Submersible Type	*166*
5.6.5	Triceratops Type	*166*
5.7	Wind Turbine on a Triceratops	*166*
5.8	Response of a Triceratops-Based Wind Turbine to Waves	*166*
5.8.1	Free-Decay Response	*166*
5.8.2	Response to Operable and Parked Conditions	*169*
5.8.3	Effect of Wave Heading Angles	*170*
5.8.4	PSD Plots	*171*
5.8.5	Tether Response and Service Life Estimation	*172*
5.9	Stiffened Triceratops	*173*
5.9.1	Preliminary Design	*173*
5.9.2	Response to Wave Action	*175*
5.9.3	Effect of Wave Direction	*177*
5.10	Triceratops with Elliptical Buoyant Legs	*179*

5.10.1 Conceptual Development *180*
5.10.2 Response of a Triceratops with Elliptical Buoyant Legs to Wave Action *182*
5.11 Summary *186*

Model Test Papers *187*
References *209*
Index *223*

List of Figures

1.1	A typical tension leg platform.	9
1.2	TLP mechanics.	10
1.3	Active control strategy.	25
1.4	Semi-active control strategy.	26
1.5	Block diagram for passive control strategy.	26
1.6	Pall friction damper.	27
1.7	Metallic yield damper.	28
1.8	Viscous fluid damper.	28
1.9	Tuned liquid damper: (a) circular; (b) rectangular.	29
1.10	Tuned liquid column damper.	30
1.11	Tuned mass damper.	32
1.12	Schematic diagram of an idealized system.	32
1.13	Schematic diagram of a spring-mass system with a TMD.	33
1.14	Mass used in the TMD.	39
1.15	Response of a TLP and TMD ($\mu = 1.5\%$).	41
1.16	Response of a TLP and TMD ($\mu = 3.0\%$).	41
1.17	Surge RAO of a TMD.	42
1.18	Comparison of surge response ($H_S = 8$ m; $T_P = 12$ seconds).	42
1.19	Comparison of surge response ($H_S = 8$ m; $T_P = 16$ seconds).	43
1.20	Comparison of surge response ($H_S = 8$ m; $T_P = 20$ seconds).	43
1.21	Comparison of surge response ($H_S = 8$ m; $T_P = 32.5$ seconds).	44
1.22	Comparison of pitch response ($H_S = 8$ m; $T_P = 12$ seconds).	45
1.23	Comparison of pitch response ($H_S = 8$ m; $T_P = 16$ seconds).	45
1.24	Comparison of pitch response ($H_S = 8$ m; $T_P = 20$ seconds).	46
1.25	Comparison of pitch response ($H_S = 8$ m; $T_P = 32.5$ seconds).	46
1.26	Articulated tower.	47
1.27	Analytical model.	49
1.28	Variation of responses for different frequency ratios.	51
1.29	Variation of responses for different frequency ratios with a TMD.	52

1.30	Variation of responses for different frequency and mass ratios with a TMD.	*52*
1.31	Geometric details of the model.	*54*
1.32	Model of a TMD.	*54*
1.33	Free-vibration time history.	*55*
1.34	Surge response of a MLAT without a TMD.	*55*
1.35	Surge RAO for TMD-1.	*56*
1.36	Surge RAO for TMD-2.	*56*
1.37	Surge RAO for TMD-3.	*57*
1.38	Comparison of RAOs for 3 cm wave height.	*57*
1.39	Comparison of RAOs for 5 cm wave height.	*58*
1.40	Comparison of RAOs for 7 cm wave height.	*58*
2.1	Schematic diagram of the BLSRP installed in a wave flume.	*64*
2.2	Experimental setup and arrangements: (a) side view; (b) hinged joint; (c) roller at the base plate for guiding the mooring line; (d) load cell; (e) ratchet mechanism for adjusting the tension; (f) plan arrangement.	*66*
2.3	Orientation of the BLSRP for the wave heading angle.	*69*
2.4	Response of the BLSRP (0°, 0.1 m wave height).	*70*
2.5	Details of the hinged joint in the numeric model.	*73*
2.6	Response of the BLSRP (30°, 15 m).	*74*
2.7	Tether tension variations in mooring lines.	*76*
2.8	Power spectral density plots of buoyant leg 1 (0°, 6 m, 10 seconds).	*78*
2.9	Power spectral density plots of the deck (0°, 6 m, 10 seconds).	*80*
2.10	Numerical model of the BLSRP (normal case).	*87*
2.11	Numerical model of the BLSRP with postulated failure.	*89*
2.12	Dynamic tether tension variation in postulated failure cases.	*91*
2.13	Mathieu stability for the BLSRP in postulated failure cases.	*92*
3.1	Typical regular wave profile (H = 2 m, T = 5 s).	*96*
3.2	PM spectrum for different sea conditions.	*99*
3.3	Two-dimensional random wave profile.	*99*
3.4	API spectrum plot for different wind velocities.	*101*
3.5	Wind-generated current velocity profile.	*102*
3.6	Service life estimation methodology.	*103*
3.7	Triceratops model.	*104*
3.8	Experimental model of a stiffened triceratops.	*105*
3.9	Plan of the triceratops.	*105*
3.10	RAOs of the deck and buoyant legs with regular waves.	*107*
3.11	Deck response given different wave heading angles.	*109*
3.12	Tether tension variation in rough sea conditions.	*110*
3.13	Deck surge and heave PSD plots in very high sea conditions.	*111*
3.14	Pitch response of the deck and buoyant legs in very high sea conditions.	*112*

3.15	Maximum deck response in very high sea conditions.	*114*
3.16	Tether tension spectrum with very high sea conditions.	*115*
3.17	Maximum tether tension in very high sea conditions.	*115*
3.18	Deck response with high sea conditions (w – waves, w + w – waves+wind, w + w + c – waves+wind+current).	*117*
3.19	Phase plots in the surge DOF with very high sea conditions.	*119*
3.20	Buoyant leg response with high sea conditions.	*120*
3.21	Tension spectrum with very high sea conditions.	*123*
4.1	Random ice force and vibration of the structure.	*128*
4.2	True stress–strain curve of AH36 grade steel.	*130*
4.3	Different shapes of indenters.	*131*
4.4	Time–temperature curves for different fire conditions.	*132*
4.5	Reduction factors for yield strength, proportional limits, and linear elastic range for carbon steel.	*133*
4.6	Variations in the thermal conductivity of carbon steel.	*133*
4.7	Variations in the specific heat of carbon steel.	*134*
4.8	Variations in the thermal strain of carbon steel.	*134*
4.9	Spectral density plot given different ice velocities.	*137*
4.10	Spectral density plot given different ice forces.	*137*
4.11	Ice force–time history.	*138*
4.12	PSD plots for normal ice sea conditions with ice load on two buoyant legs.	*141*
4.13	PSD plots of tether tension variation in normal sea conditions.	*143*
4.14	Total deck response for different ice thicknesses.	*144*
4.15	Total deck response for different ice crushing strengths.	*145*
4.16	Total deck response for different ice velocities.	*146*
4.17	PSD plots of the deck in open water and ice-covered load cases.	*147*
4.18	Methodology of impact analysis.	*148*
4.19	Numerical model of buoyant legs and indenters.	*149*
4.20	Force versus nondimensional deformation curve.	*150*
4.21	Deck surge responses for impact loads on buoyant leg 1.	*150*
4.22	Force–deformation curves for different indenter sizes.	*151*
4.23	Force–deformation curves for different impact locations.	*152*
4.24	Force–deformation curves for different indenter shapes.	*153*
4.25	Force–deformation curves for different numbers of stringers.	*154*
4.26	Force–deformation curve of buoyant legs at different temperatures.	*155*
4.27	Deck plate of a triceratops.	*156*
4.28	Scale deck plate model.	*157*
4.29	Hydrocarbon fire cases.	*158*
4.30	Temperature variations in plates and stiffeners.	*159*
5.1	Numerical model of a triceratops with a wind turbine.	*168*
5.2	Pitch RAO of the triceratops.	*168*

5.3	PSD plot of the surge free-decay response.	*169*
5.4	PSD plot of the roll free-decay response.	*169*
5.5	Frequency response to operable and parked conditions.	*170*
5.6	PSD plots for different DOF.	*172*
5.7	Dynamic tether tension variation.	*172*
5.8	Plan and elevation of a stiffened buoyant leg.	*174*
5.9	Fabricated model of a stiffened buoyant leg.	*176*
5.10	Fabricated model of a ball joint.	*177*
5.11	Surge, heave, and pitch RAOs of the deck and buoyant legs with 0° incident waves.	*178*
5.12	Surge, heave, and pitch RAOs of the deck and buoyant legs with 90° waves.	*179*
5.13	Surge, heave, and pitch RAOs of the deck and buoyant legs with 180° waves.	*180*
5.14	Effect of wave direction on the stiffened triceratops.	*181*
5.15	Cross section of the buoyant legs.	*182*
5.16	Plan view of the triceratops with circular and elliptical buoyant legs.	*183*
5.17	Total force–time history, given high sea conditions.	*185*

List of Tables

1.1 Major fixed platforms constructed worldwide (as of 2017). *3*
1.2 Tension leg platforms constructed worldwide. *20*
1.3 Properties of a TMD in the model and prototype (scale 1 : 100). *39*
1.4 Results of free oscillation tests of the TLP model. *40*
1.5 RMS value of surge responses in the presence of random waves. *44*
1.6 RMS value of pitch responses in the presence of random waves. *47*
1.7 Mechanical properties of the Perspex material used for the model. *53*
2.1 Structural details of the BLSRP. *67*
2.2 Natural periods and damping ratios. *68*
2.3 Maximum response of the BLSRP model (0°, 0.1 m). *72*
2.4 Maximum response amplitudes (numerical studies; 6 m wave height). *83*
2.5 Geometric properties of the BLSRP for the stability study. *88*
2.6 Maximum tension amplitude in the tethers in postulated failure cases. *90*
2.7 Mathieu parameters in postulated failure cases. *92*
2.8 Fatigue life (rounded off) of tethers under eccentric loading. *93*
3.1 Characteristics of random sea conditions. *97*
3.2 Comparison of responses to regular waves. *108*
3.3 Deck response to different sea conditions. *110*
3.4 Comparison of deck responses to high sea conditions. *112*
3.5 Tension variation and service life of tethers of buoyant leg 1. *115*
3.6 Characteristics of sea conditions. *116*
3.7 Tether tension variation with combined actions of wind, waves, and current. *122*
4.1 Mechanical properties of marine DH36 steel. *131*
4.2 Ice sea conditions. *138*
4.3 Deck response to different sea conditions. *139*
4.4 Deck response to open water and ice-covered load cases. *146*
4.5 Collision speed and impact duration. *148*
4.6 Mechanical properties of DH36 steel at a 0.001/s strain rate. *155*

- 5.1 Properties of the triceratops-based wind turbine. *167*
- 5.2 Variation in RAO with changes in the wave heading angle. *171*
- 5.3 Service life estimation of the triceratops. *173*
- 5.4 Geometric parameters of the offshore triceratops. *175*
- 5.5 Mass properties of the triceratops. *176*
- 5.6 Response of the triceratops given rough sea conditions. *184*

Foreword by Professor Purnendu K. Das

Advances in technology and industry maturity make offshore wind an increasingly attractive investment. Although still relatively expensive, it has advantages of being deployable sooner and faster than many other nonrenewable energy sources. Compared to other renewable sources, offshore wind turbine technology has advantages in scalability. Recent growth and innovation have driven costs to more competitive levels and significant future investments in Europe and globally. Current drawbacks include high capital costs due to the large fabrication, installation, and maintenance costs involved; it is estimated that over 20% of total project costs are directly linked to the foundation structures and their construction.

The book describes the detailed analysis and design procedures of compliant offshore structures with a special focus on new-generation platforms like the triceratops and buoyant leg storage and re-gasification platforms. The book aims to describe the detailed preliminary design of a triceratops in ultra-deep water. A detailed analysis under environmental loads that are inherent in offshore locations, such as waves, wind, and currents, is presented. A new methodology for the dynamic analysis of a triceratops under ice loads, predominantly in ice-covered regions, is also explained, with detailed parametric studies. Because offshore platforms are also prone to accidental loads arising due to fires and ship–platform collisions, the detailed dynamic analysis under such loads discussed in the book will be of great assistance to both researchers and practicing structural consultants.

I hope this book will serve as a ready reference for engineers in this field who want to study floating wind turbines structures. I wish the book all success.

Professor Purnendu K. Das B.E., M.E., Ph.D., C.Eng, CMar.Eng, FRINA, FIStruct.E, FIMarEST
Director
ASRANet Ltd.
Glasgow
Ex-Professor of Marine Structures, University of Strathclyde, Glasgow, UK
Visiting Professor, University of Montenegro, Montenegro

Foreword by Dr. Atmanand N.D.

The use of renewable energies is vital for addressing issues due to global warming and climate change. But the cost of production of renewable energy has not hit an all-time low as yet, and oil and gas continue to be the major sources of energy.

In the recent past, offshore oil drilling and production platforms have begun moving toward ultra-deep water due to the depletion of oil and gas resources near shore. In addition, the arctic region is opening for new offshore platforms. This necessitates a novel geometric form with reduced response to extreme waves and, in turn, the extreme loading conditions that prevail in ultra-deep water. Compliant offshore platforms are highly popular due to their form--dominant design characteristics. However, their significant hull motion in deepwater conditions and high sea conditions leads to a need for alternate design procedures, because the present ones are not suitable for ultra-deep water. Detailed analysis and design procedures for new-generation offshore platforms are frequently debated in conference proceedings. But this book demystifies the technological know-how by presenting a lucid explanation that is useful and innovative. For example, the discussion of a new methodology for the dynamic analysis of a triceratops under ice loads in ice-covered regions, with detailed parametric studies, is noteworthy. Such structures are prone to accidental loads arising due to fires and ship–platform collisions, so both designers and researchers should be familiar with the detailed dynamic analysis under such loads. The comprehensive picture presented in this book of the dynamic response behavior of this novel platform under different types of loads is scarce elsewhere in the literature.

This book will serve as a resource for understanding the basic structural behavior of new-generation complex offshore platforms and will help graduate students understand analysis methodologies that otherwise would have to be painstakingly collected from many publications. In addition, this book will be useful for practicing engineers and research scholars who wish to understand the response behavior of structures with novel geometry under combinations of extreme loads.

The principal author, Srinivasan Chandrasekaran, is a well-known academician and has authored 14 textbooks in the highly specialized area of offshore engineering. His web-based courses on offshore structural engineering are very popular and serve as reference material to teach this complex subject at both the undergraduate and post-graduate levels of engineering programs in various disciplines including civil, mechanical, aerospace, naval architecture, etc. I am sure any offshore engineer will find this book to be a wealth of resources.

Dr. Atmanand N.D.
Director, Chair IOCINDIO (UNESCO)
National Institute of Ocean Technology
Chennai
India

Series Preface

The Wiley-ASME Press Series in Mechanical Engineering brings together two established leaders in mechanical engineering publishing to deliver high-quality, peer-reviewed books covering topics of current interest to engineers and researchers worldwide.

The series publishes across the breadth of mechanical engineering, comprising research, design and development, and manufacturing. It includes monographs, references and course texts.

Prospective topics include emerging and advanced technologies in Engineering Design; Computer-Aided Design; Energy Conversion & Resources; Heat Transfer; Manufacturing & Processing; Systems & Devices; Renewable Energy; Robotics; and Biotechnology.

Preface

This book, *Offshore Compliant Platforms: Analysis, Design, and Experimental Studies,* describes detailed analysis and design procedures for compliant offshore structures, with a special focus on new-generation platforms like the triceratops and buoyant leg storage and regasification platforms. While the conceptual development of conventional platforms like tension leg platforms (TLPs), spar platforms, and articulated towers is presented briefly, the detailed descriptions of the design and development of new-generation platforms discussed in the book are highly novel and still in the preliminary stages of study in the existing literature.

Compliant offshore platforms are favorable candidates for deepwater oil and gas production systems due to their form-dominant design characteristics. But significant compliancy causing flexible motion in the horizontal plane requires special attention from designers because it poses critical challenges when platforms are commissioned in ultra-deep water. Therefore, a novel geometric form with reduced responses is a vital necessity to accommodate extreme loading.

This book presents a detailed analysis and design of one such novel platform: the triceratops. The authors believe that it will serve as a good reference guide for the effective design of triceratops platforms, as the clear numerical and experimental studies presented in the book will help readers understand the platforms' dynamic response behavior. A new methodology for the dynamic analysis of a triceratops under ice loads in ice-covered regions is also explained with detailed parametric studies. Offshore platforms are also prone to accidental loads arising due to fires and ship–platform collisions; the detailed dynamic analysis under such loads that is presented in the book will be of great interest to both researchers and practicing structural consultants.

In addition, this book will aid in understanding the platform's structural behavior in terms of its response, service life, and design. The book will serve as a resource regarding the basic structural behavior of complex offshore structures; it will help graduate students understand analysis methodologies and will also help researchers understand the dynamic response of such structures. Readers will learn about new structural geometries of offshore platforms and different methods of analysis for assessing their performance under special loads. The discussion of fatigue analysis and predicting service life will also help professionals during the preliminary and detailed design stages of offshore platforms. This book can serve as reference material for both academicians and offshore practicing professionals.

Both senior undergraduate and post-graduate students in the disciplines of civil, mechanical, aerospace, structural, offshore, and ocean engineering; applied mechanics; and naval architecture will find this book very useful as a standard classroom reference for analysis and design of special structures. In addition, this book will be useful for practicing engineers and research scholars studying the response behaviors of structures with novel geometry under combinations of extreme loads.

The experimental studies and numerical analyses discussed in the book are the outcomes of research work carried out recently by the authors and research scholars supervised by Srinivasan Chandrasekaran. All discussions, interpretations, and concepts conceived during the detailed research work carried out by the research scholar team are sincerely acknowledged. Administrative support extended by the Centre for Continuing Education (CCE), Indian Institute of Technology, Madras in preparing this manuscript is sincerely acknowledged.

Srinivasan Chandrasekaran
R. Nagavinothini

1

Common Compliant Platforms

Summary

This chapter presents details of the structural geometry of compliant offshore platforms while emphasizing the design and development of these platforms rather than of fixed platforms. Details of investigations carried out on the response control of a tension leg platform (TLP) and an articulated tower are presented. This chapter also presents various control strategies that are commonly deployed for response control of structures subjected to lateral loads while presenting detailed applications of tuned mass dampers (TMDs) in offshore structures. Experimental investigations and numerical analyses, reported by R. Ranjani (2015, "Response Control of Tension Leg Platform Using Tuned Mass Damper," PhD thesis, IIT Madras, India) and Prof. Deepak Kumar (Dept. of Ocean Engineering, IIT Madras, India) are sincerely acknowledged by the authors.

1.1 Introduction

While it is a common understanding that structural forms are conceived to counteract the applied loads acting on them, it may not be completely true in the context of offshore structures. This is due to the basic fact that offshore structures are designed to alleviate the encountered environmental loads based on their geometric form and not by the strength of the materials and structural properties based on the cross-section of the members. On the other hand, offshore structures are essentially form-dominant. Therefore, rigid-supported structural systems, which are conventionally good to transfer the applied loads to the foundations, are not attractive candidates for offshore structures. A few factors that influence

---

Offshore Compliant Platforms: Analysis, Design, and Experimental Studies,
First Edition. Srinivasan Chandrasekaran and R. Nagavinothini.
© 2020 John Wiley & Sons Ltd.
This Work is a co-publication between John Wiley & Sons Ltd and ASME Press.

the choice of the geometric form of offshore structures are a structural form with a stable configuration, and a geometric form with low installation, fabrication, and decommissioning costs; that leads to lower capital expenditures (CAPEX) and a higher return on investment (ROI). The result is an earlier start for production and greater mobility, and a platform that requires the least intervention so that uninterrupted production can take place.

Offshore structures are grouped as fixed, compliant, and floating types, based on the boundary conditions of their station-keeping characteristics. Fixed platforms are further categorized as jacket platforms, gravity platforms, and jack-up rigs. Compliant platforms are further categorized as guyed towers, articulated towers, and tension leg platform (TLPs). Floating platforms are further categorized as semi-submersibles, floating production units (FPUs), floating storage and offloading (FSO), floating production storage and offloading (FPSO), and spar. While fixed platforms are designed to remain insensitive to lateral loads that arise from waves, wind, currents, and earthquakes, these platforms are expected to resist the encountered loads using both their material strength and structural redundancy. A strength-based design approach is more popular for use when designing such platforms, as they exhibit very low displacement under lateral loads. These structural forms are also stable due to their massive weight, which is an advantageous design factor. However, with regard to the other factors that influence the choice of the structural forms of offshore structures, as explained earlier, fixed platforms are least preferred for deepwater construction. Table 1.1 shows a list of fixed platforms constructed worldwide (Srinivasan Chandrasekaran 2017).

It is interesting to note that offshore engineers gradually realized that restricting the motion of the platform by making it fixed to the seabed was not necessary. Instead, conceptual developments were focused on flexible structural forms and gave birth to compliant structures. Further, fixed platforms became increasingly expensive and difficult to commission in greater water depths. Hence, a modified design concept was evolved in offshore structural forms, which focused on flexible systems; it gave rise to compliant offshore structures. The word *compliance* refers to flexible systems. Fixed platforms were designed to encounter the environmental loads with their strength and fixity (rigidness, obtained from both large member dimensions and fixity to the seabed).

On the other hand, compliant-type offshore platforms are allowed to undergo large displacements but are position-restrained using cables or tethers. In addition to the material strength that helps to counter environmental loads, a major contribution came from the relative displacement concept of the design: compliant offshore platforms do not resist loads based on their fixity to the seabed but rather receive lesser loads due to their motion characteristics.

Table 1.1 Major fixed platforms constructed worldwide (as of 2017).

S. no	Platform name	Water depth (m)	Location
	North America		
1	East Breaks 110	213	USA
2	GB 236	209	USA
3	Corral	190	USA
4	EW910-Platform A	168	USA
5	Virgo	345	USA
6	Bud Lite	84	USA
7	Falcon Nest	119	USA
8	South Timbalier 301	101	USA
9	Ellen	81	USA
10	Elly	81	USA
11	Eureka	213	USA
12	Harmony	365	USA
13	Heritage	328	USA
14	Hondo	259	USA
15	Enchilada	215	USA
16	Salsa	211	USA
17	Cognac	312	USA
18	Pompano	393	USA
19	Bullwinkle	412	USA
20	Canyon Station	91	USA
21	Amberjack	314	USA
22	Bushwood	***	USA
23	Hebron	92	Canada
24	Hibernia	80	Canada
25	Alma	67	Canada
26	North Triumph	76	Canada
27	South Venture	23	Canada
28	Thebaud	30	Canada
29	Venture	23	Canada
30	Ku-Maloob-Zaap (KMZ)	100	Mexico

(*Continued*)

Table 1.1 (Continued)

S. no	Platform name	Water depth (m)	Location
	South America		
1	Peregrino Wellhead A	120	Brazil
2	Hibiscus	158	Trinidad and Tobago
3	Poinsettia	158	Trinidad and Tobago
4	Dolphin	198	Trinidad and Tobago
5	Mahogany	87	Trinidad and Tobago
6	Savonette	88	Trinidad and Tobago
7	Albacora-Leste	***	Peru
	Australia		
1	Reindeer	56	Australia
2	Yolla	80	Australia
3	West Tuna	***	Australia
4	Stag	49	Australia
5	Cliff Head	**	Australia
6	Harriet Bravo	24	Australia
7	Blacktip	50	Australia
8	Bayu-Undan	80	Australia
9	Tiro Tiro Moana	102	New Zealand
10	Lago	200	Australia
11	Pluto	85	Australia
12	Wheatstone	200	Australia
13	Kupe	35	New Zealand
	Asia		
1	QHD 32-6	20	China
2	Penglai	23	China
3	Mumbai High	61	India
4	KG-8	109	India
5	Bua Ban	***	Thailand
6	Bualuang	60	Thailand
7	Arthit	80	Thailand
8	Dai Hung Fixed Well head	110	Vietnam
9	Ca Ngu Vang	56	Vietnam

Table 1.1 (Continued)

S. no	Platform name	Water depth (m)	Location
10	Chim Sao	115	Vietnam
11	Oyong	45	Indonesia
12	Kambuna	40	Indonesia
13	Gajah Baru	***	Indonesia
14	West Belumut	61	Malaysia
15	Bukha	90	Oman
16	West Bukha	90	Oman
17	Al Shaheen	70	Qatar
18	Dolphin	***	Qatar
19	Zakum Central	24	UAE
20	Mubarek	61	UAE
21	Sakhalin-I	***	Russia
22	Lunskoye-A	48	Russia
23	Molikpaq	30	Russia
24	Piltun-Astokhskoye-B	30	Russia
25	LSP-1	13	Russia
26	LSP-2	13	Russia
27	Gunashli Drilling and Production	175	Azerbaijan
28	Central Azeri	120	Azerbaijan
29	Chirag PDQ	170	Azerbaijan
30	Chirag-1	120	Azerbaijan
31	East Azeri	150	Azerbaijan
32	West Azeri	118	Azerbaijan
33	Shah Deniz Production	105	Azerbaijan
	Europe		
1	Brage	140	Norway
2	Oseberg A	100	Norway
3	Oseberg B	***	Norway
4	Oseberg C	***	Norway
5	Oseberg D	100	Norway
6	Oseberg South	100	Norway
7	Gullfaks A	138	Norway

(*Continued*)

Table 1.1 (Continued)

S. no	Platform name	Water depth (m)	Location
8	Gullfacks B	143	Norway
9	Gullfacks C	143	Norway
10	Sleipner A	80	Norway
11	Sleipner B	**	Norway
12	Sleipner C	***	Norway
13	Valhall	70	Norway
14	Ekofisk Center	75	Norway
15	Varg Wellhead	84	Norway
16	Hyperlink	303	Norway
17	Draugen	250	Norway
18	Statfjord A	150	Norway
19	Statfjord B	***	Norway
20	Statfjord C	290	Norway
21	Beatrice Bravo	290	UK
22	Jacky	40	UK
23	Ula	40	UK
24	Inde AC	70	UK
25	Armada	23	UK
26	Auk A	88	UK
27	Fulmar A	84	UK
28	Clipper South	81	UK
29	Clair	24	UK
30	Brae East-I	140	UK
31	Lomond	113	UK
32	Brae East-II	86	UK
33	Alwyn North A	126	UK
34	Alwyn North B	126	UK
35	Cormorant Alpha	126	UK
36	Dunbar	145	UK
37	Nelson	***	UK
38	Schooner	100	UK
39	Andrew	117	UK

Table 1.1 (Continued)

S. no	Platform name	Water depth (m)	Location
40	Forties Alpha	107	UK
41	Forties Bravo	107	UK
42	Forties Charlie	107	UK
43	Forties Delta	107	UK
44	Forties Echo	107	UK
45	Eider	159	UK
46	Elgin	93	UK
47	Elgin PUQ	93	UK
48	Franklin	93	UK
49	Babbage	42	UK
50	Alba North	158	UK
51	Alba South	138	UK
52	Judy	80	UK
53	Amethyst	30	UK
54	Buzzard	100	UK
55	Brigantine BG	29	UK
56	Brigantine BR	***	UK
57	Cecilie	60	Denmark
58	Nini East	***	Denmark
59	Nini	58	Denmark
60	South Arne	60	Denmark
61	Galata	34	Bulgaria

Compliant offshore platforms are position-restrained by cables or tethers. A high pre-axial loaded tether offers resistance to lateral loads while counteracting the large buoyancy force. Therefore, in the strength-based design, the concept is shifted toward a displacement-controlled design. Large displacements allowed in the design demand a recentering ability for the platform. *Elasticity* refers to material characteristics and ensures that a member regains its form, shape, and size when the applied load is within the elastic limit and is removed, and *recentering* is an extension of this property. Recentering refers to the capability of the structural form (not a material characteristic) to regain its initial position (which may not be an equilibrium position) in the presence of external forces (not when they are removed, unlike elasticity). This is very important in the context of compliant platform design, because large displacements are permitted as a part of the design. Apart from ensuring the attainment of the original position, recentering ensures the safety of other existing appurtenances and auxiliary fixtures like risers, pipelines, electric cables, and umbilicals that are connected to the compliant platform.

One of the fundamental differences between elasticity and recentering is that the former is a material characteristic, while the latter is a geometric characteristic. Design of the geometric form of compliant structures is principally dominated by balancing the buoyancy force and weight of the platform. While a preferred design is to make the buoyancy force exceed the weight, because this can enable easy installation, the difference between the two forces is balanced by an external axial force on the tethers (or cables) as pre-tension. Large displacements of compliant platforms invoke the participation of the added mass of the body through a variable submergence effect. It helps restore the dynamic equilibrium in the system in the presence of various environmental loads. One of the most successful structural forms of compliant offshore platforms is the TLP.

1.2 Tension Leg Platforms

Several TLP design concepts have been developed as the result of research and studies done on various elements of TLPs, which makes them suitable for different environments. TLPs operate successfully in deep waters and are installed at depths of a few thousand meters; they are classified as hybrid compliant structures that are suitable for both drilling and production operations.

A conventional TLP has four legs, called *columns*. The four columns are connected with a pontoon ring at the base of the column, as shown in Figure 1.1. A square or rectangular deck is placed at the top of the columns. The TLP is held in position with the help of taut, moored, pre-tensioned tethers that connect the platform columns to the seabed through the pile templates. Due to the excess buoyancy of the TLP, the tethers are always in pre-tension. A structure can have six degrees of freedom (DOF): three translation motions and three rotational motions. TLPs are designed in a way that tension in the tethers allows horizontal motion but restricts vertical motion. Hence TLPs are flexible in the surge, sway, and yaw DOF, while the heave, pitch, and roll DOF are restrained. Thus TLPs have high natural periods in the surge, sway, and yaw DOF, and low natural periods in the heave, pitch, and roll DOF. TLPs are designed such that the natural periods are far separated from the wave frequency band. The heave natural period is reduced by increasing the pipe wall thickness of the tethers, and the natural pitch period is decreased by the spacing of the tethers.

Figure 1.1 A typical tension leg platform.

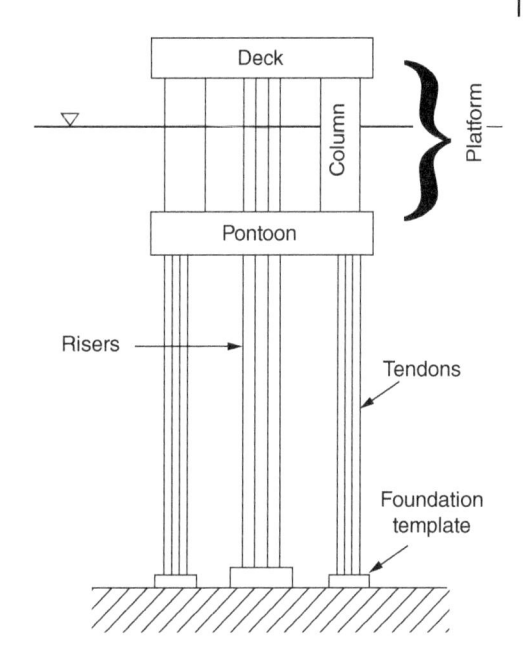

When no load is applied to the structure, the structure is in a stationary, stable condition. Due to excess buoyancy, the tethers are in high tension, so the platform is held down to the seabed. Given lateral loading from wind, waves, or currents, the structure experiences a lateral displacement. This lateral displacement of the TLP is called *offset*. The offset condition of the TLP pulls down the structure. The vertical displacement is called *setdown*. Since the tethers are in high tension, the setdown effect is minimal. The horizontal component of the pre-tension in the tethers produces a restoring force to the structure, bringing the hull back to its original position. There is a tension variation, which is due to the variable submergence effect. This flexibility of the TLP in horizontal DOF creates nonlinearity in the stiffness of the structure due to large displacements. This phenomenon affects the dynamics of the entire structural system. A schematic diagram of TLP setdown is shown in Figure 1.2. During this phenomenon, there are changes in the tension of the tethers. Arbitrary displacement of the TLP produces dynamic tether-tension variation.

Compliancy of TLPs, introduced in the design of the structural form, enables the effective negotiation of encountered environmental loads (Adrezin et al. 1996; El-gamal et al. 2013). Analytical studies carried out on TLPs showed that high elasticity of tethers controls the motion response of TLPs through a lesser variation in tether tension (Bar-Avi 1999; Booton et al. 1987; Mekha et al. 1996). Methods for motion response analyses were addressed, and the method for the

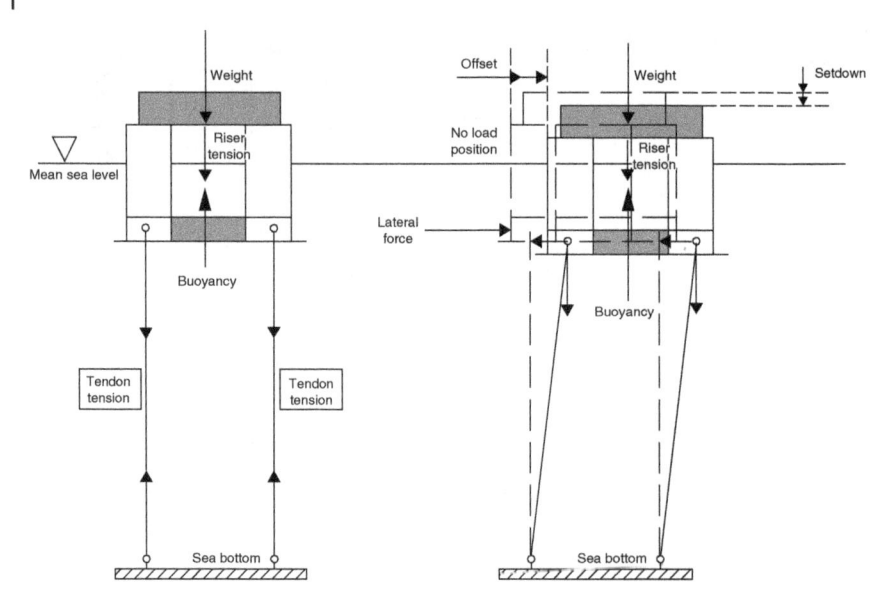

Figure 1.2 TLP mechanics.

structural design was also discussed based on fatigue analysis and other design procedures (Chandrasekaran and Jain 2002a, b, 2004; Chandrasekaran et al. 2004, 2007a, b, 2011). The influence of variable submergence on added mass terms and the coefficients of the stiffness matrix were addressed in addition to the stability analysis (Chandrasekaran et al. 2006b, 2007c; Kim et al. 2007). The results were compared with those of an eight-legged jacket structure and guyed tower by highlighting the advantages of TLP.

Jefferys and Patel (1982) studied the Mathieu instability of a TLP in sway motion using an energy-balance technique. Predicted sway motions were within the acceptable limits even in the presence of large waves, which were caused by Mathieu excitation (Donely and Spanos 1991; Ertas and Lee 1989; Ertas and Eskwaro-Osire 1991). The results showed that square law variation of fluid damping influenced the upper bound of the amplitude of oscillation. Yoneya and Yoshida (1982) carried out experimental investigations on the dynamic response of a TLP to regular waves in a wave tank. Numerical analysis was carried out on a full-scale model in real sea conditions to extrapolate the response behavior of the prototype under operational sea conditions. The analytical methods were validated and compared with the experimental results. Patel and Lynch (1983) developed a mathematical model to study the coupled dynamics of a tensioned buoyant surface platform with taut tethers. The results show that the platform response is influenced by tether dynamics with long tethers in comparison to the water depth and payload, respectively (Chandrasekaran and Gaurav 2008). de Boom et al. (1984) carried out experiments on scale models with both regular and irregular waves. Time-domain based potential theory was used to predict motion and tether forces on a TLP. The results of the analytical and experimental studies were in good agreement except for a few discrepancies at low frequencies. The reasons were attributed to the estimation of the wave in the presence of drift forces, and improper modeling of the second harmonics of tether tension given regular waves at low frequencies (Chou 1980; Gasim et al. 2008; Ramachandran et al. 2014). Smiu and Leigh (1984) discussed the analysis methodology to estimate the surge response to turbulent wind in the presence of current and waves; nonlinearities arising from hydrodynamic forces were included in the analysis. An alternate wind spectrum with ordinate at the origin is used by several researchers to account for the fluctuations in wind speed at very low frequencies (Jain and Srinivasan Chandrasekaran 2004; Kurian et al. 2008; 1993; Leonard and Young 1985; Logan et al. 1996). The results show that given extreme wave conditions, significant wind-induced resonant amplification effects occurred even for small drag coefficient in the Morison equation (Jain 1997; Kobayashi et al. 1987; O'Kane et al. 2002).

Spanos and Agarwal (1984) studied the response of a TLP model given wave forces at the displaced position of the platform by modeling a TLP as a single degree of freedom (SDOF) system. The Morison equation was used to predict wave forces on the structure, and the equation of motion was solved by the numerical integration method. It was shown that the proposed simplified method remains valid for both deterministic and stochastic wave loads. Yoshida et al. (1984) presented the equations for the linear response analysis method with regular waves to find the response motion. Tension variations of tendons and structural member forces were solved for the structural response of a TLP, and this method was confirmed by comparison with the test results on two small-scale TLP models simultaneously. Leonard and Young (1985) presented the coupled response of compliant offshore structures with examples of articulated towers, guyed towers, and TLPs. Three-dimensional finite element analysis (3D FEA) was used to simulate the static and dynamic coupled responses. It was concluded that the procedure used in the study requires a longer time for simulation in the case of a TLP using the Morison equation to predict the wave forces for all of the considered compliant structures (Jefferys and Patel 1982; Mercier et al. 1997; Muren et al. 1996).

Booton et al. (1987) and Yashima (1976) conducted a parametric study to estimate the effect of tether damage on a TLP by reducing tether in the presence of regular waves at a water depth of 160 m without changing the pre-tension. Nonlinear analytical results also showed that the amplitudes of the motion were more than the linear analyses results the lateral motion of the platform was not influenced by the loss of tether stiffness but increased heave response was observed (Perryman et al. 1995; Vickery 1990). Kobayashi et al. (1987) performed a study on the dynamic response of a TLP to random waves; experiments and analytical studies were performed with time and frequency domains using three-dimensional singularity distribution methods to determine the wave forces and concluded that the time domain analytical results compared well with the experimental results. Nordgren (1987) studied the dynamic response of a TLP using spectral analysis closer to natural heave, pitch, and roll periods of vertical DOF of the exciting waves; resonant response and the fatigue life of tethers were examined. They concluded that the resonant response is reduced by the combined effect of radiation damping and material damping of tethers.

Ertas and Lee (1989) carried out a stochastic analysis of a TLP given random waves in the presence of constant current velocity. A modified Morison equation was used to estimate wave forces considering the relative motion in both the drag and inertia terms, while the superposition method was used for random waves in the time domain. The responses of a TLP in surge DOF with and without current were compared, and the conclusion was that the frequency domain analysis technique was quite efficient. Vickery (1990) studied the response of a TLP in the presence of the combined wind and wave loads; experimental studies were conducted on 1 : 200 scale model in a wind-wave flume with scaled wind and wave loads. First- and second-order wave loads were estimated using the Morison equation and diffraction theory in the numerical model, respectively. Wind loads were estimated using quasi-static theory (Davenport 1961). The equation of motion was solved using the fourth-order Runge–Kutta differential equation solver in the time domain. Based on the studies, it was concluded that the surge response due to wind loads dominates the overall response of a TLP.

Roitman et al. (1992) reported the experimental responses of a small-scale model of a TLP in deep water and compared the results with the numerical model. The results confirmed that the study of numerical simulations for the effects of impact loads and wave loads are performed well. Kurian et al. (1993) conducted experimental studies on a scaled TLP in the presence of regular and random waves with different wave directions. They showed that TLPs exhibited a satisfactory response control and were suitable for deepwater applications (Low 2009; Yan et al. 2009; Younis et al. 2001). Adrezin et al. (1996) reviewed the dynamic response of different compliant offshore structures with the highlights of various modeling approaches of TLPs in the presence of wind, wave, and current forces.

Mekha et al. (1996) performed a nonlinear coupled analysis to study the implication of tendon modeling on the response of a TLP. The tendons were modeled as mass-less elastic springs that were connected to the hull. The responses of TLPs in different water depths were studied, and the results also showed that tether tension variation was on the order of 10% of the initial pre-tension value (Tabeshpour et al. 2006; Thiagarajan and Troesch 1998; Vannucci 1996). It was concluded that the lateral stiffness of the TLP should be modeled carefully to predict the lateral response. Logan et al. (1996) conducted a feasibility study on a three-column concrete mini TLP for marginal deepwater fields and showed that it is technically and economically superior as compared to other floating platforms. Alternative concepts such as one-column TLPs, two-column TLPs, and three-column TLPs with steel truss pontoons were considered for the study; they reported that a three-column TLP is the optimum solution for marginal fields (Zeng et al. 2007). Murray and Mercier (1996) performed hydrodynamic tests on 1 : 25 scale model of an Ursa TLP with truncated tendons. The study highlighted the importance of model tests for predicting the responses of deepwater TLPs.

Muren et al. (1996) presented a three-column TLP design for 800 m water depth, which is easily extended to 1500 m with minimal time and cost. The dynamic response was computed using a radiation-diffraction code augmented with linearized Morison elements; model tests at 1 : 100 scale were carried out to verify the numerical prediction. Vannucci (1996) proposed a simplified method to design an optimal TLP depending on two variables, such as effective structure and dimensions of the hull, to obtain required buoyancy. The results agreed well with the available data. Jain (1997) performed a nonlinear time-history analysis using Newmark's method to study the coupled response of offshore TLPs in the presence of regular waves. Airy wave theory was used to estimate water particle kinematics, and the Morison equation was used to estimate the wave forces by neglecting diffraction effects. Mercier et al. (1997) conducted tests on a scale model of a Mars TLP in the presence of waves, wind, and current loads. A 1 : 55 scale ratio was chosen for towing tests that were carried out at three different drafts to ensure floatation stability, while hydrodynamic tests were carried out on a 1 : 200 scale model. Numerical analysis was performed to validate the experimental results, which were subsequently considered in the global design of the Mars TLP. Chakrabarti (1998) addressed different models' similitude, techniques, deepwater testing requirements, environments, and areas of testing. Thiagarajan and Troesch (1998) conducted model tests to examine the effects on TLP columns in the presence of waves with uniform current. The results showed that the heave damping induced due to the disk was linear for the amplitude of oscillation. Bar-Avi (1999) studied the response of a TLP given various environmental loads such as wind, wave, seismic force, and current by considering geometric and external force nonlinearities. Based on the studies, the importance of such analyses was highlighted for the commissioning of the structure. Buchner et al. (1999) combined the requirements of water depth, wave, wind and current generation, hydraulic design, and wave absorption in a deepwater offshore basin to conduct experiments on deepwater and ultra-deepwater structures. Niedzweki et al. (2000) estimated surface wave interactions with a compliant deepwater mini-TLP and two spar platforms by using analytical expressions and a series of model tests. The Weibull distribution was employed on the prediction of the response histogram, and the analytical results compared well with the experimental results. Younis et al. (2001) proposed fundamental equations that govern the turbulent flow based on computational fluid dynamics and an alternate method to estimate hydrodynamic forces on a full-scale mini TLP. The results obtained from both methods were compared with the experimental data observed on 1 : 70 scale models, and they showed good agreement for the front column members of a TLP. Stansberg et al. (2002) reviewed the challenges in verifying hydrodynamic forces computed on deepwater structures. Model tests with truncated moorings were carried out, and the results were compared with those obtained using computer simulations and other hybrid approaches. The results highlighted the complexities and limitations of each of the investigation methods in addition to the future challenges associated with them.

Chandrasekaran and Jain (2002a) compared the dynamic behavior of square and triangular TLPs given regular wave loading. The results show that a triangular TLP exhibits a lower response in the surge and heave DOF than that of a four-legged (square) TLP. It was also reported that the pitch response of a triangular TLP was more than that of a four-legged (square) TLP. Chandrasekaran and Jain (2002b) investigated the dynamic response of a triangular TLP in the presence of random sea waves. The random sea waves were generated using the Pierson-Moskowitz (PM) spectrum with different significant wave heights and peak periods. Airy theory was used for water particle kinematics. Morison theory was used for wave force estimation, and the equation of motion was solved using the Newmark-beta method. The effect of current was also investigated, with the conclusion that the coupling responses of the structure were greater given the combined effect of wave and current. O'Kane et al. (2002) proposed a new method for estimating the added mass and hydrodynamic damping coefficients of a TLP. Model tests were conducted on one, two, and four columns of a TLP, with the conclusion that based on the Keulegan–Carpenter number, a TLP with multiple columns behaves identically to a single-column TLP with respect to the damping ratio estimates.

Bhattacharyya et al. (2003) reported on the coupled dynamics of a SeaStar mini TLP using Morison type wave loading at two water depths: 215 and 1000 m. Experimental investigations were performed for a scale model corresponding to 215 m water depth for validation of the numerical model. The response amplitude operators (RAOs) of the SeaStar at two water depths were compared to highlight limitations related to the experimental investigations by indicating the influence of water depth on the response of a TLP. Liagre and Niedzwecki (2003) estimated the coupled responses of a TLP in the time and frequency domains by considering nonlinear effects such as nonlinear stiffness, inertia, and damping forces. They verified the numerical results obtained from WAMIT with those of the experimental measurements. Chandrasekaran et al. (2004) studied the influence of hydrodynamic drag and inertia coefficients on the response behavior of triangular

TLPs in the presence of regular waves. The response of two triangular TLPs at 600 and 1200 m water depths were compared; drag and inertia coefficients were assumed based on the Reynolds and Keulegan–Carpenter numbers. Wave forces were estimated using the modified Morison equation while neglecting diffraction effects. Based on the studies, it was concluded that the influence of hydrodynamic coefficients was greater in the case of a 15 second wave period than a 10 second wave period.

Jain and Chandrasekaran (2004) performed numerical analyses to study the aerodynamic behavior of a triangular TLP due to low-frequency wind force and random waves. They included various nonlinear effects such as variable submergence, hydrodynamic force, tether tension, etc. in the study. Wave forces were estimated using the Morison equation and Airy wave theory, while wind force was estimated using Emil Simiu's wind velocity spectrum. Based on the study, it was concluded that low-frequency wind force alters the response of a triangular TLP significantly. Stansberg et al. (2004) suggested advanced numerical tools for model studies associated with ultra-deepwater structures that arise from the limitations caused by water depth in existing wave tanks. A model-the-model procedure and the final prototype simulations were presented. Chandrasekaran et al. (2006a) investigated the seismic analysis of a triangular TLP in the presence of moderate regular waves by considering nonlinearities due to the change in tether tension and nonlinear hydrodynamic drag forces. El Centro and Kanai-Tajimi's earthquake data were considered for the analysis, and the vertical ground displacement was imposed as the tether tension variation. Based on the studies, it was concluded that the TLP heave response was affected and found to vary nonlinearly. Chandrasekaran et al. (2006b) conducted the numerical studies on TLPs; the water depth increases with deepwater-compliant structures, and tether tension variation plays an important role in the stability analysis of TLPs. Lower pre-tension values, as shown in the Auger TLP, lead to instability. Triangular TLPs with three groups of tethers are more stable than four-legged TLPs in the first fundamental mode of vibration.

Tabeshpour et al. (2006) developed a computer program called STATELP using MATLAB to perform a nonlinear dynamic response analysis of a TLP in both the time and frequency domains given random sea wave loading. The PM spectrum was used to generate the random waves; the modified Morison equation and Airy theory were used to predict the wave loads. The power spectral density (PSD) functions of displacement, velocity, and acceleration were calculated from the nonlinear responses in different DOF, and the safety of personnel and facilities on board were examined. Xiaohong et al. (2006) developed a numerical code called COUPLE to study the dynamic response of TLPs, including the tendons and risers. The comparisons of experimental and numerical results showed that wave loads on a mini TLP were accurately predicted using the Morison equation and also concluded that COUPLE was able to predict the dynamic interaction between the hull and its tendon and riser systems while the related quasi-static analysis fails. Bas Buchner and Tim Bunnik (2007) investigated the effect of extreme wave loads on the response of floating structures resulting from the air gap, green water on the deck, and slamming to the hull. Green-water effects were studied on floating FPSOs, and dynamic response studies were studied on a TLP using experimental and time integration analytical investigations. The improved volume of fluid method was adopted in the time domain analysis to predict the green-water effect. Based on the studies, it was concluded that extreme waves could damage the floating structure due to the air gap, green water on the deck, or slamming to the hull. The experimental and analytical predictions compared well. Chandrasekaran et al. (2007a) presented the response behavior of triangular TLPs in the presence of regular waves using Stokes fifth-order nonlinear and Airy wave theories. The wave forces were estimated using the modified Morison equation. The drag and inertia coefficients were varied along with the depth of the TLPs. The results showed that the responses were higher when Airy wave theory was considered than with Stokes wave theory. Chandrasekaran et al. (2007b) performed a dynamic analysis of two different triangular TLPs at 527.8 and 1200 m water depths with 45° inclined impact load and wave loads. The impact load was considered as triangular and half-triangular loads. The wave force was estimated using the modified Morison equation and Stokes fifth-order wave theory. The results indicated that the impact load significantly affected the response when it acted on the columns rather than the pontoons.

Chandrasekaran et al. (2007c) studied the response behavior of triangular TLPs using the dynamic Morison equation considering nonlinearities associated with vortex shedding effects, variable submergence, variable added mass, stiffness, damping, and variable hydrodynamic coefficients along with the depth of the columns. The wave force was estimated using Stokes fifth-order nonlinear wave theory. Based on the studies, it was concluded that the triangular TLP responses were higher in deeper water, and the dynamic Morison equation was capable of estimating vortex shedding effects on the columns. Cheul-Hyun Kim et al. (2007) developed a numerical code to study the response of a TLP in the presence of regular waves by assuming the TLP as a flexible model. Source distribution and radiation effects were considered in the analysis. The numerical results compared well with the established experimental and analytical results. Zeng et al. (2007) modeled an International Ship Structures Congress (ISSC) TLP for its displacement, mass, and tether tension. The equation of motion was solved in the time domain by considering nonlinearities such as instantaneous wet surface, submerged volume, displacement, velocity, and acceleration of the structure. The responses were compared with the published results and verified well. Chandrasekaran and Gaurav (2008) performed an earthquake motion analysis on three different triangular TLPs in the presence of high sea waves. Seismic excitation was generated using the Kanai-Tajimi earthquake spectrum. The results showed that the tension variation of the tethers given the combined loads was higher than the regulation values, and TLPs in deeper water showed lesser responses. Gasim et al. (2008) developed MATLAB coding to study the dynamic response of square and triangular TLPs in the presence of random waves. Airy wave theory and the Morison equation were used to estimate wave force. A time history analysis was performed using the Newmark-beta method, and RAOs of a square TLP were compared with the established results to validate the MATLAB code. The study was extended to triangular TLPs, and the RAOs of the square and triangular TLPs were compared. The results showed that except for surge RAO, all the responses of triangular TLPs were higher than those of square TLPs.

Kurian et al. (2008) developed a MATLAB program to study the nonlinear response of a square TLP at 300 m water depth in the presence of regular and random waves. Airy wave theory and the Morison equation were used to determining the wave loading. The Newmark-beta method was used to solve the equation of motion. The numerical results were extended to 600 m water depth and predicted higher responses at 600 m water depth. The RAOs compared well with the available experimental and theoretical results. Low (2009) performed a frequency domain analysis of a TLP by linearizing the system. Wave forces were computed using WAMIT. Four separate cases with different static offsets were studied. The linearization technique agreed well with the time domain results with a few limitations. Yan et al. (2009) compared the experimental and analytical stress response amplitudes of a TLP under extreme environmental loads. The analytical predictions were done in WAMIT. Strain gauges were placed on the model to measure the strain record of the platform. Stress RAOs were plotted using a fast Fourier transform technique, and the experimental and analytical results were compared. Jayalekshmi et al. (2010) presented a nonlinear program using Lagrangian coordinates and Newmark integration methods to investigate the effect of tether-riser dynamics on the response of deepwater TLPs at 900 and 1800 m water depths in the presence of random waves. Linear wave theory and the Morison equation were used to estimate the wave forces; current forces were also considered in the analysis. Based on the results, it was concluded that the response of the TLP was high at 1800 m water depth. Chandrasekaran et al. (2010) stated that ringing waves are highly nonlinear and contain strongly asymmetric transient waves; hence the shape of such waves becomes critical in offshore compliant structures and is presented along with the method of analysis. Frequency responses were observed to be almost the same for all cases, indicating that all platforms showed ringing as the peaks are not observed near the corresponding natural frequencies. This study develops a mathematical formulation of the impact waves responsible for ringing and examines their influence on offshore TLPs with different geometric configurations located at different water depths.

Chandrasekaran et al. (2011) developed a mathematical formulation of impact and non-impact waves and discussed the method of analysis for triangular TLPs. Their effects are more comparable to square TLPs. Impact and non-impact waves are responsible for ringing and springing phenomena, respectively. Ringing (caused by impact waves in the pitch DOF) and springing (caused by non-impact waves in the heave DOF) in both platform geometries are undesirable because they present a serious threat to platform stability. Analytical studies show that equivalent triangular TLPs positioned at different water depths are less sensitive to these undesirable responses, thus making them a safe alternative for deepwater oil explorations.

El-Gamal et al. (2013) presented numerical studies for square TLPs using the modified Morison equation; they were carried out in the time domain with water particle kinematics using Airy linear wave theory to investigate the effect of changing the tether tension force on the TLP. The Newmark-beta integration method was used to solve the nonlinear equation. The surge response showed high-amplitude oscillations that were significantly dependent on wave height, and indicated that special attention should be given to tether fatigue because of the tethers' high tensile static and dynamic stress. Chandrasekaran et al. (2013) quantified the response variations of a triceratops – a new-generation offshore platform that alleviates wave loads with its innovative structural form and design. In the presence of seismic activities, the proposed new-generation offshore platforms are very useful. While they are positively buoyant, the platform is position-restrained by tethers and ball joints. Ball joints connect the deck and the buoyant leg structure (BLS) units and also restrain the transfer of rotations between them. These platforms experience significant tether-tension variation given vertical seismic excitations caused at the seabed (Ney Roitman et al. 1992; Chandrasekaran et al. 2006a). Records of the El Centro earthquake and the artificially generated earthquake using the Kanai-Tajimi (K-T) power spectrum are considered for the study. Chandrasekaran et al. (2013) develop a mathematical formulation for the aerodynamic analysis of an offshore triceratops and examines its response in the presence of regular waves and wind. Based on the numerical studies conducted here, it is observed that the triceratops shows a significant reduction in deck response, with no transfer of rotation from the BLS; the deck remains horizontal given the encountered wave loads. The response of the deck is relatively lower than in the BLS units in the pitch and yaw DOF even at higher significant wave heights. The geometric form is advantageous for keeping more facilities on the deck system and operating even given moderate weather conditions. Chandrasekaran et al. (2015) develop an offshore triceratops. Experimental studies are performed to study the dynamic response of the triceratops in coupled DOFs given regular waves with a unidirectional wave on the scale model. Higher natural periods in the surge, sway, and heave DOF of the triceratops indicate a higher degree of compliance in comparison to other compliant type offshore platforms like TLPs. Numerical studies are also carried to verify the experimental results and dynamic responses of the triceratops in the presence of different wave headings for regular and random waves. Based on the detailed investigations carried out, it is seen that the coupled responses of the deck in the rotational DOF are less than those of the buoyant legs.

Ramachandran et al. (2014) implemented a three-dimensional fully coupled hydro-aero-elastic model for a floating TLP wind turbine with 17 DOF. The aerodynamic loads were implemented using unsteady blade element momentum theory, which includes the effect of a moving tower shadow, wind shear using a power law, and spatially coherent turbulence, whereas the hydrodynamic loads were implemented using the Morison equation. Loads and coupled responses were predicted for a set of load cases with different wave headings. An advanced aero-elastic code, Flex5, was extended for the TLP wind turbine configuration, and comparing the response with the simpler model showed generally good agreement, except for the yaw motion. This deviation was found to be a result of missing lateral tower flexibility in the simpler model.

Table 1.2 summarizes the TLPs commissioned worldwide (Chandrasekaran 2017). TLPs have proved their efficiency for deepwater oil exploration in terms of stability and cost-efficiency.

1.3 Guyed Tower and Articulated Tower

Guyed towers are compliant offshore structures whose position restraint is ensured by guy wires. Guyed towers are generally used for small field production and installed up to a water depth of about 500 m. Guy wires limit the lateral motion of the platform deck, while pile and clump weights are used to position the guyed towers on the seabed. Given extreme sea conditions, clump weights are raised from the seafloor, providing an additional restoring force to the tower. Articulated towers are similar to TLPs, with tethers replaced by a single high-buoyancy shell. The buoyant shell offers the required restoring force to counteract encountered lateral loads. The universal joint is a unique component of the tower, which connects it to the foundation system. Depending on the nature of the seabed at the site, the foundation system may consist of a concrete base or a pile foundation. Articulated towers are generally installed at a water depth of about 500 m.

Witz et al. (1986) performed model tests and theoretical studies on rigid semi-submersibles and conducted experimental studies on articulated semisubmersibles. Articulations were placed at the base of the columns and pinned to the vessel base and maintained excess buoyancy over self-weight. Comparison of righting moment curves indicated that the articulated structure showed higher righting

Table 1.2 Tension leg platforms constructed worldwide.

S. no	Platform name	Water depth (m)	Location
		USA	
1	Shenzi	1333	USA
2	Auger	872	USA
3	Matterhorn	869	USA
4	Mars	896	USA
5	Marlin	986	USA
6	Brutus	1036	USA
7	Magnolia	1433	USA
8	Marco Polo	1311	USA
9	Ram Powell	980	USA
10	Prince	454	USA
11	Neptune	1295	USA
12	Ursa	1222	USA
13	Morpeth	518	USA
14	Allegheny	1005	USA
15	Jolliet	542	USA
		Europe	
1	Snorre A	350	Norway
2	Heidrun	351	Norway
		Africa	
1	Okume/Ebano	500	Equatorial Guinea
2	Oveng	280	Equatorial Guinea

moments than the rigid structure. Experimental and analytical results agreed well, while the articulated structure showed lesser responses when compared with the rigid structure. Sellers and Niedzwecki (1992) derived the equation of motion, which was valid for both single- and multi-articulated towers using the Lagrange equation approach. A deepwater tension-restrained articulated tower was studied, and its dimensionless parameters were derived from avoiding resonance excitation given environmental loads. Bar-Avi and Benaroya (1996) presented studies of the response of an offshore articulated tower subjected to deterministic random wave loading. The tower was modeled as an upright rigid pendulum mass concentrated at the top and hinged at the base with Coulomb friction and viscous structural damping. The equation of motion was solved using Borgman's method considering the wave height, wave frequency, buoyancy, drag geometric, and wave force nonlinearities. The wave force was estimated using the modified Morison equation. The PM spectrum was used to generate random waves. Parametric studies were also performed, such as buoyancy, initial condition, wave height, frequency, current velocity, and direction. Based on the studies, it was observed that Coulomb damping reduced the beating phenomenon and RAO.

Nagamani and Ganapathy (2000) studied the dynamic response of a three-legged articulated tower by conducting an experiment and analysis. The legs of the articulated tower were connected to the seabed with universal joints, and the deck and legs were connected with a ball-and-socket joint. It was addressed that the articulated structure should have the stability characteristics of less acceleration in the deck and the smallest possible loading on the articulated joint. The effects of mass distributions on the variations of bending moment and deck acceleration were discussed. Based on the study, it was concluded that the experimental and analytical results compared well, and the bending moments and deck acceleration increased with an increase in wave height. Islam and Ahmad (2003) investigated the relative importance of the seismic response of an articulated offshore tower in comparison to the response due to wave forces. The equation of motion was derived using the Lagrangian approach, and the solution was obtained with the Newmark-beta integration scheme. It included nonlinearities associated with variable submergence, drag force, coulomb damping, variable buoyancy, and added mass, along with the geometrical nonlinearities of the system and the joint occurrence of the waves and seismic forces together with the current given random sea conditions.

1.4 Floating Structures

A floating production system (FPS) is a floating unit that is fully equipped with exploration and production equipment. While in operation, the unit is positioned at the site using either anchors or rotating thrusters. Hydrocarbons explored from the subsea system are subsequently transported to the surface with the help of production risers. FPSs are deployed at water depths ranging from 600 to 2500 m. The FPSO system is a big tanker-type ship that is positioned on the seabed with the help of either anchors or dynamic positioning systems. A FPSO is designed to process and store the oil produced from nearby subsea wells as well. Stored oil is periodically offloaded to a smaller shuttle tanker, which transports the oil to an onshore facility for further processing. A FPSO may be suited for a marginally economic field located in remote deepwater areas where a pipeline infrastructure is not economically feasible.

With further advancements in the process of conceiving more appropriate structural forms to make platforms insensitive to water depth, floating structures have been introduced. A spar platform is a large, deep-draft, cylindrical floating caisson, generally used for exploration and production purposes, and installed at water depths of a few thousand meters. The spar has a long cylindrical shell called a *hard tank*, which is located near the water level. It generates high buoyancy to the structure that helps stabilize the platform; the midsection is annular and used for free flooding. The bottom part is called a *soft tank* and is utilized for placing the fixed ballast. It essentially floats the structure during transport and installation. In order to reduce the weight, drag, and cost of the structure, the midsection is designed to be a truss structure. To reduce the heave response, horizontal plates are introduced between the truss bays. The cell spar is the third generation of spar platforms, which was commissioned in 2004. It has several ring-stiffened tubes that are connected by horizontal and vertical plates. The hull is transported to the offshore site horizontally on its side. At the desired location, the structure is ballasted at the required attitude and then installed. The topside is attached at the site once the installation is complete.

Newman (1963) developed a linearized potential wave theory to estimate the regular wave forces on a spar buoy and presented the response amplitude in the surge, pitch DOF. A spar buoy was modeled separately as an undamped and damped system. Response amplitudes and phase angle plots of damped and undamped systems were discussed. Glanville et al. (1991) analyzed a spar floating drilling production storage structure in the presence of the combined effects of wave, wind, and current loads at a water depth of 800 m. The authors highlighted the stability characteristics, fabrication, and ease of installation of the spar. The equation of motion, which included mass, stiffness, and linearized damping matrices, was solved using a time integration scheme. The results indicated that the spar was cost-effective, insensitive to payloads, and best suited for deepwater drilling and production applications (Agarwal and Jain 2002; Finn et al. 2003; Glanville et al. 1991).

Ran et al. (1996) performed experimental and analytical investigations on a 1 : 55 scale model given regular, bichromatic, and unidirectional irregular waves, with and without sheared currents. Time-domain numerical investigations were performed on the spar in the presence of waves and currents; the wave-structure interaction was modeled considering second-order wave theory, including the effects of viscous and wave-drift damping. Bichromatic and bidirectional wave forces were estimated using second-order diffraction/radiation methods, and the Morison drag formula was used to estimate viscous drag forces. The experimental and numerical results were in good agreement, and the spar responses were within the practically acceptable limits in 100-year-storm sea conditions (Koo et al. 2004). Agarwal and Jain (2002) performed a time-history analysis of a spar platform under wave loads using Airy wave theory, the Morison equation, and the Newmark-beta method considering a spar as a system with six DOF. The spar mode considered the effect of mooring lines connected at fairlead locations. Heave, pitch, and roll stiffness were modeled based on hydrostatics, while the lateral stiffness was modeled using nonlinear horizontal springs. Based on the studies, it was concluded that the inertia and drag coefficients influence the surge and heave responses of the platform (Montasir et al. 2008; Montasir and Kurian 2011).

Finn et al. (2003) performed vortex-induced vibration studies on a cell spar in the presence of waves and current loads in different wave directions. The model was fabricated with different configurations of strakes, and the tests were conducted on the cell spar model, with and without strakes. The model with strakes reported a lower sway response in comparison to that without strakes, and an optimum strake configuration was also suggested. Koo et al. (2004) highlighted the causes of Mathieu instability for a spar due to variable submergence and time-varying metacentric heights in the coupled heave and pitch motions. Effects of the hull, mooring, and riser coupling on the principal instability and damping effects were studied using a modified Mathieu equation, and the effect of wave elevation was also investigated. Based on the simulations, it was concluded that Mathieu instability increased with the increase in pitch motion. Damping was found to suppress the instability, and the additional pitch-restoring moment due to buoyancy also contributed to a reduction in Mathieu instability.

Zhang et al. (2007) performed a numerical simulation on a spar in operating and survival sea conditions. They investigated a cell spar moored with nine mooring lines in a chain-wire-chain form in three groups. Vertical pre-tension in the mooring lines was assumed in the simulation. The Joint North Sea Wave Project (JONSWAP) and Norwegian Petroleum Directorate (NPD) wind spectra were considered and performed in time-domain simulations. Based on the studies, it was concluded that the motions of the spar were found satisfactorily; but the low-frequency motions required care because resonance in heave motions was present in horizontal motions, and second-order drift motions were present in vertical motions. Montasir et al. (2008) performed a dynamic analysis of classic and truss spars in the presence of unidirectional regular and random waves. The spar was considered a rigid body and connected to the sea floor by catenary mooring lines. The water particle kinematics were estimated using Chakrabarti's stretching formula, and wave forces were determined based on the modified Morison equation. Based on a time-history analysis using the Newmark-beta method, the results showed that the spar had better motion characteristics and was economical. Montasir and Kurian (2011) developed a MATLAB code called TRSPAR to analyze a truss spar with strakes in the presence of slowly varying drift forces, and validated it with experimental results. Slowly varying frequency wave forces were derived, and the Morison equation was used for force estimation. The developed code considered various nonlinearities like variable submergence, nonlinear springs for mooring lines, and drag. Neeraj Aggarwal et al. (2015) discussed an offshore wind turbine installed on a spar platform at a water depth of 320 m. The coupled wind and wave analysis was achieved by coupling the FAST aerodynamic software and ANSYS AQWA hydrodynamic software. The power spectral densities of the response were obtained by using the transfer function of the system for irregular sea conditions defined by the Pierson-Moskowitz (PM) spectrum, where the wave parameter was chosen near rough sea conditions ($Hs = 6$ m, $Tp = 10$ s). The Gumbel method shows lower values compared to extremes estimated using the Weibull method. The sensitivity to sample size was also not significant.

As seen from the discussion, offshore structures with different geometric forms are becoming increasingly common and have large displacements under lateral loads. The degree of compliance imposed by their design makes them suitable for deepwater and ultra-deepwater exploration. However, large displacements also make them unsuitable for safe operability. The ideal situation is to have an offshore platform that is highly compliant but does not undergo large displacement under lateral loads, which seems to be entirely hypothetical. However, a few recent studies have attempted to control the large displacements of the deck without compromising their compliancy, as discussed in the following section.

1.5 Response Control Strategies

Recent trends in the geometric design of several structural forms focus on the use of lightweight, highly durable materials. This has increased the probability of the development of structural configurations that are elastic and low damping. A *damper* is a device used to restrain, depress, or reduce the motion of a structure. Dampers are efficient at controlling energy input due to dynamic loading through various mechanisms. The energy gained due to dynamic loading is dissipated through the various mechanisms as either heat or deformation.

Unlike land-based structures, offshore structures under environmental conditions are subjected to secondary vibrations that initiate structural failure, discomfort during topside activities, and breakdown of equipment. Response control methods with passive dampers and base-isolation systems are among the few successful applications on land-based structures but cannot be directly deployed in offshore structures for several reasons. One of the most important and complicated tasks is to tune the frequency of the system vibration to that of the external control devices, such as dampers. Dampers are activated by the motion of the structure and dissipate energy via various mechanisms. Secondary damping devices, which are commonly useful in response control, can be classified into three groups: active, semi-active, and passive systems.

1.5.1 Active Control Algorithm

Active systems, once installed, continuously monitor structural behavior. After processing the information over a short time, they generate a set of forces to modify the current state of the structure. A block diagram of the active control strategy is shown in Figure 1.3. An active control system consists of three major components: (i) the monitoring system, which perceives the present state of the structure and subsequently records the data using an electronic data-acquisition system; (ii) the control system, which determines the reaction forces to be applied to the structure based on the communication received from the monitoring system; and (iii) the actuating system, which applies the physical forces to the structure, as directed by the control system. To accomplish all these things, an active control system needs a continuous external power source. The loss of power that might be experienced during a catastrophic event may render these systems ineffective. A few common examples of this kind are active mass dampers and active liquid dampers.

1.5.2 Semi-Active Control Algorithm

Semi-active systems are similar to active systems except that they need less external power for successful activation. Semi-active control devices are also often viewed as controllable passive devices. Instead of exerting additional forces on the

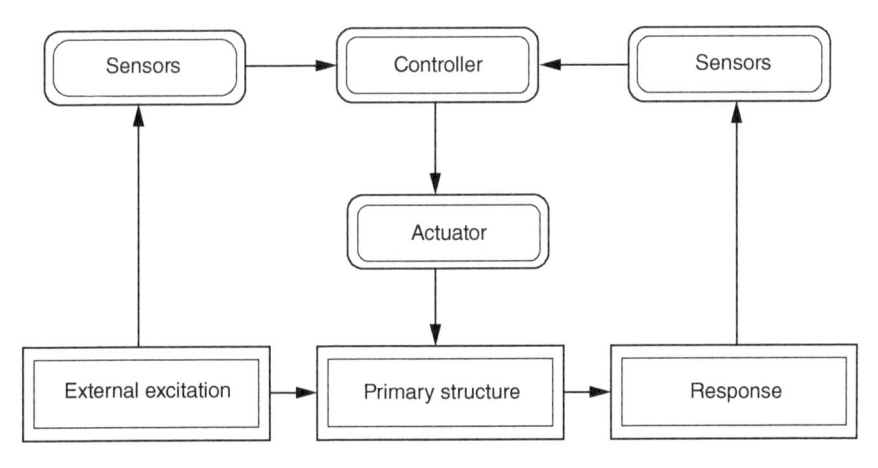

Figure 1.3 Active control strategy.

26 Offshore Compliant Platforms

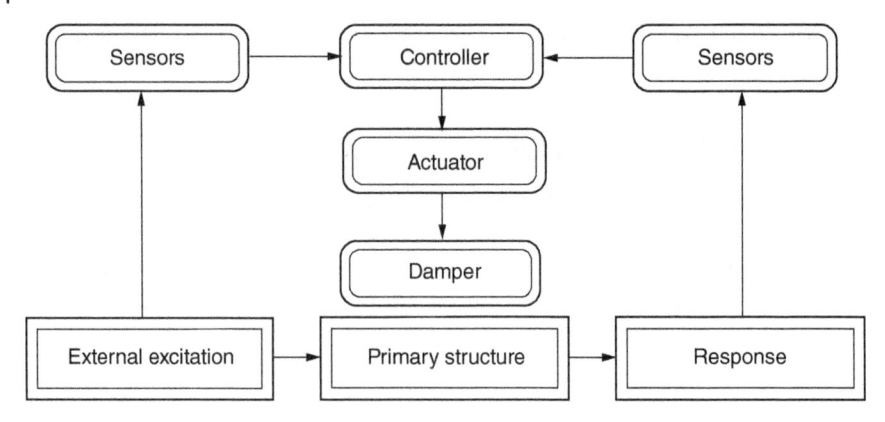

Figure 1.4 Semi-active control strategy.

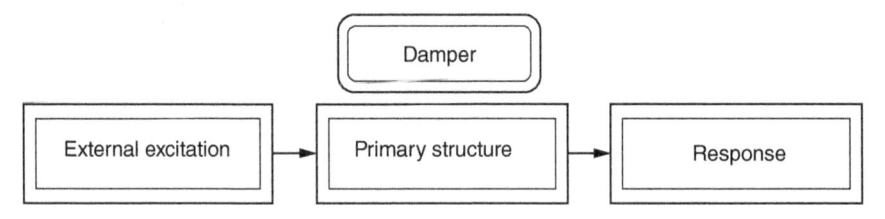

Figure 1.5 Block diagram for passive control strategy.

structural systems, they control vibrations by modifying the structural characteristics. A block diagram representing a semi-active control system is shown in Figure 1.4. The need for an external power source to make these systems functional limits their application in structural engineering, as they may fail due to catastrophic events. Examples of semi-active devices include variable orifice fluid dampers, controllable friction devices, variable stiffness devices, controllable fluid dampers, and magneto-rheological dampers.

1.5.3 Passive Control Algorithm

Passive systems require no external energy for successful operation, which is one of the major advantages of such systems in comparison to the former types. A key benefit of passive control devices is that once installed in a structure, they do not require any startup or operation energy, unlike active and semi-active systems. A block diagram representing a passive control system is shown in Figure 1.5. Passive control devices are active at all times until maintenance, replacement, or dismantling is required. Passive control systems include friction dampers, metallic yield dampers, and viscous fluid dampers. Alternative types of passive control systems contain a spring (or spring-like component), which is tuned to a particular natural frequency of the structure for maximum damping. Examples of these passive control devices are tuned mass dampers (TMDs), tuned liquid dampers (TLDs), and tuned liquid column dampers (TLCDs).

1.5.4 Friction Dampers

Friction dampers are devices installed at the connection of cross-bracing, introduced by Pall in the Canadian Space Agency (Pall et al. 1993). A friction damper consists of two solid bodies that are compressed together. It uses the friction between the two surfaces to dissipate energy. As a structure is subjected to vibration, the two bodies slide against each other, developing friction that dissipates the energy of the motion. These devices have been built into structures and have been successful in providing enhanced seismic protection by being designed to yield during extreme seismic vibration. Wind loads do not provide enough shear force to activate these types of dampers. Figure 1.6 shows a Pall frictional damper.

1.5.5 Metallic Yield Dampers

The typical design for this damper is a triangular or X-shaped plate that absorbs vibrations through the inelastic deformation of the metallic material. These devices are known to have stable hysteretic behavior. They are usually installed in newly built and retrofitted buildings and are successful in reducing seismic loads. Figure 1.7 shows a metallic yield damper.

1.5.6 Viscous Fluid Dampers

Viscous fluid dampers are similar to conventional shock-absorbers. They consist of a closed cylinder-piston, which is filled with fluid (usually silicon oil). The

Figure 1.6 Pall friction damper.

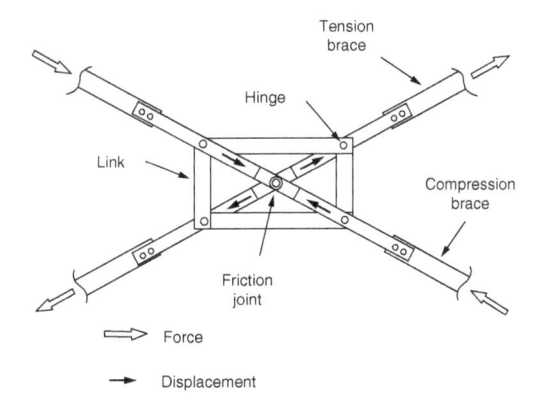

28 | *Offshore Compliant Platforms*

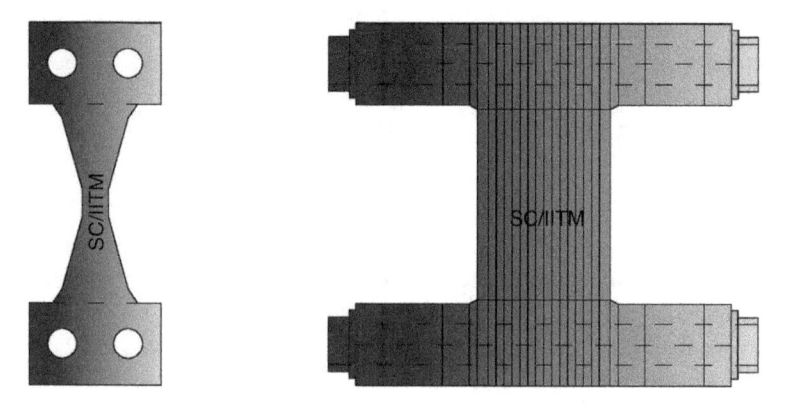

Figure 1.7 Metallic yield damper.

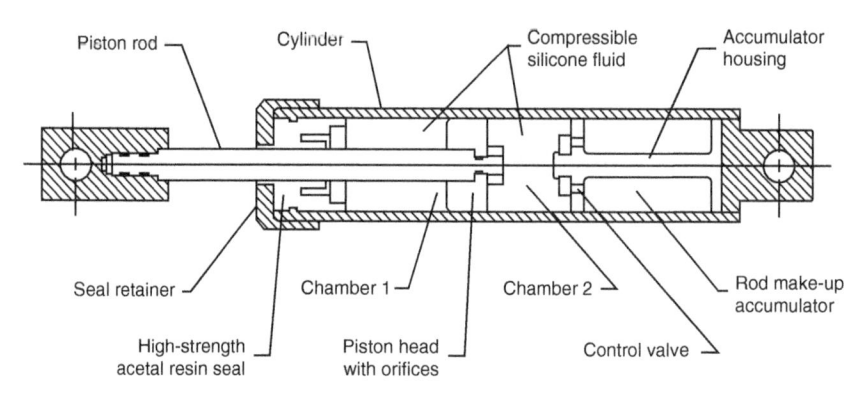

Figure 1.8 Viscous fluid damper.

detailed structural elements of the viscous damper are shown in Figure 1.8. The piston-head contains orifices that regulate the flow of fluid between the two chambers of the piston. When the structure is excited, movement of the fluid through the holes generates friction, and subsequently heat, which dissipates the motion of the structure. These dampers are typically installed as diagonal braces in building frames (preferably steel structures). To provide optimal damping, buildings are often equipped with multiple dampers in place of the diagonal beams on every floor. These viscous dampers have proved to be effective in controlling the vibrations of slender structures under lateral loads (Samuele Infanti et al. 2008).

1.5.7 Tuned Liquid Dampers

TLDs are either rectangular or circular and are installed at the highest floor of a building to control the maximum displacement. Figure 1.9 shows the configurations of TLDs. Depending on the height of the water in the tank, a TLD is categorized as a shallow-water or deepwater TLD (Ahsan Kareem 1990). If the ratio of the height of the liquid column in the damper to the length of the tank (in the case of a rectangular tank) or diameter of the circular tank is lesser than 0.15, then it is classified as a shallow-water TLD (Kareem and Sun 1987). It is important to note that the height of the liquid column in the damper depends on the natural frequency of the structure to be controlled. When the frequency of the tank motion is closer to one of the natural frequencies of the tank fluid, large-amplitude sloshing occurs. Tuning the TLD parameters to the structural parameters induces high sloshing and wave-breaking phenomenon. It helps to dissipate significant energy from the primary structure, resulting in the reduction of structural vibrations.

Ahsan Kareem (1990) suggests a new approach to reduce the building response using TLDs. Structural analyses of the building in response to wind loads were investigated. The boundary layers near the tank walls are assumed to dissipate all the energy. The sloshing in the TLD reduces the structure response by increasing the damping to the system. The vibration of the system is absorbed and dissipated due to wave-breaking and viscous effects. The response reduction by introducing the damper to the structure is effective when the TLD is tuned to the structure frequency. Sun et al. (1995) proposed a nonlinear analytical model of a rectangular tank with a shallow-water TLCD for pitch motion using shallow-water wave theory. The damping of liquid sloshing affects efficiency, and hence it is a significant parameter of the study. The results of the proposed analytical model show a considerable reduction in the resonant pitch response of the primary structure. The effectiveness of mitigating pitch motion depends on the liquid mass in the container, the configuration, and the location of the TLD

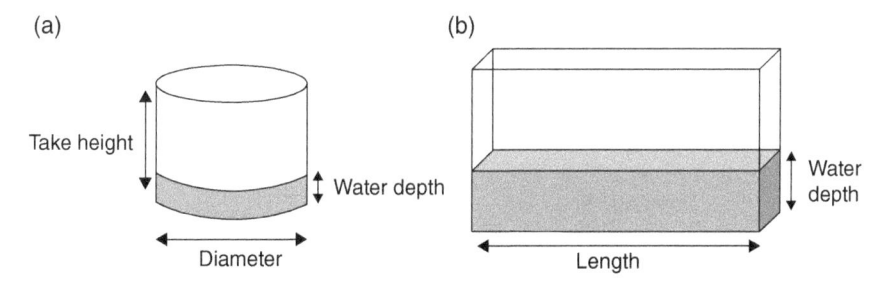

Figure 1.9 Tuned liquid damper: (a) circular; (b) rectangular.

on the structure. Analytical and experimental studies prove that the liquid sloshing in the rectangular tank is nonlinear given a pitching motion. Tait et al. (2008) experimentally investigated the performance of both unidirectional and bidirectional TLDs in response to random excitation. The performance of the TLD was examined under various loads, tuning ratios, and ratios of water depth to tank length. The parametric studies resulted in the development of performance charts, based on shallow-water wave theory, which could help design TLDs for unknown structural frequency.

1.5.8 Tuned Liquid Column Damper

A TLCD is a U-shaped tube half-filled with liquid, as seen in Figure 1.10. Unlike a TLD, which depends on liquid sloshing to dampen structural vibrations, a TLCD controls structural motion by a combined action of the movement of liquid in the tube and the loss of pressure due to the orifice inside the tube (Gao et al. 1997). A nozzle is placed at the horizontal part of the tube. The extent of response control achieved by a TLCD depends on the frequency of the exciting force acting on the structure (Fahim Sadek et al. 1998). While the restoring force is developed by the gravitational force acting on the liquid, the orifice is the controlling element for the dynamics of the liquid sloshing inside the tube. Damping depends on the opening and the type of orifice used.

Balendra et al. (1995) studied the effectiveness of TLCDs for reducing wind-induced motion of towers. The nonlinear equation is linearized to obtain the response of the tower. The Harris spectrum is used to model the along-wind turbulence. The response of the tower is modeled as a SDOF system. By conducting parametric studies, the optimum parameters for a greater response reduction are presented. A similar reduction can be obtained by using a proper opening ratio of the orifice of the TLCD. The opening ratio is varied from 0.5 to 1. It is found that a TLCD with a higher width-to-liquid-length ratio and a high

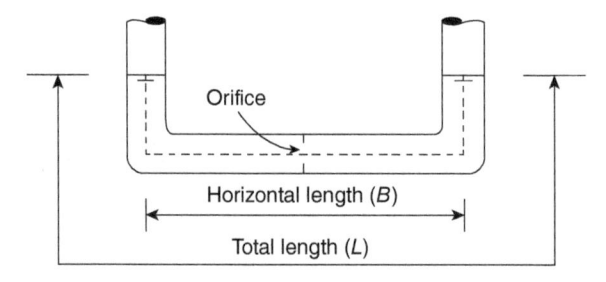

Figure 1.10 Tuned liquid column damper.

mass ratio is better for maximum reduction in acceleration and displacement. For the best response reduction, it is necessary to tune the liquid column damper to the structural frequency. If the TLCD cannot be tuned to the frequency of the structure, satisfactory control can be obtained by choosing a proper opening ratio. Jong Cheng Wu et al. (2005) proposed some useful guidelines for designing a TLCD for a damped SDOF structure. The design table provides the list of necessary optimal parameters and the corresponding response reduction for the design. An empirical formula is proposed for predicting the basic properties of TLCDs and the head-loss coefficient. The optimal tuning ratio and the head-loss coefficient are numerically obtained by minimizing the normalized response of a damped SDOF structure equipped with a TLCD under a white-noise type of wind loading. It is proved that a uniform cross-section is best for a given mass ratio and horizontal length ratio.

Anoushirvan Farshidianfar et al. (2009) studied the vibration behavior of a structure with a TLCD. Using the Lagrange equation, an unsteady and non-uniform flow equation for the TLCD is investigated. By using the normalized mean square value of the nondimensional structure response as the performance index, the analytical formulas of the optimum TLCD parameters for the undamped structure are derived. The tuning frequency ratio, length ratio, and mass ratio are obtained by using the white-noise type of wind excitation, and the performance of the TLCD for controlling the wind-induced responses of a 75-story flexible skyscraper is investigated. For a given length and damping ratio, the optimal value of the tuning frequency ratio is close to 1. Increasing the length ratio can give better control. A mass ratio greater than 3% is impossible to use and does not guarantee good control. A TLCD with a uniform cross-section is recommended for optimum response reduction.

1.6   Tuned Mass Dampers

A TMD is a passive type of damper that imposes response control using the principle of inertia. It applies indirect damping to the structural system. The inertial force of the damper is made to be equal and opposite the excitation force for optimum control. TMDs are used for structures under lateral loads. Figure 1.11 shows a schematic diagram of a TMD. It consists of a secondary mass attached to the main structure through a spring-dashpot arrangement. The energy of the primary structure is dissipated by the inertial force produced by the damper. The damper produces an inertial force in the direction opposite the direction of motion of the structure. The inertial force, in the opposite direction, helps reduce the motion of the primary structure. For maximum response reduction, the parameters of the TMD need to be tuned with

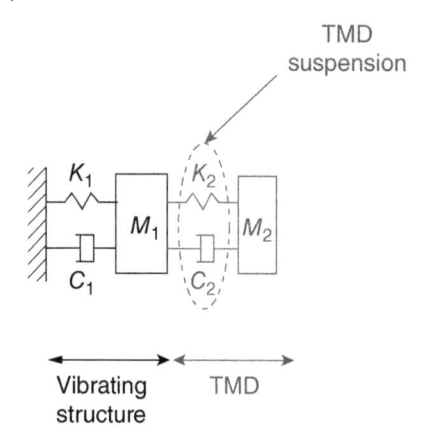

Figure 1.11 Tuned mass damper.

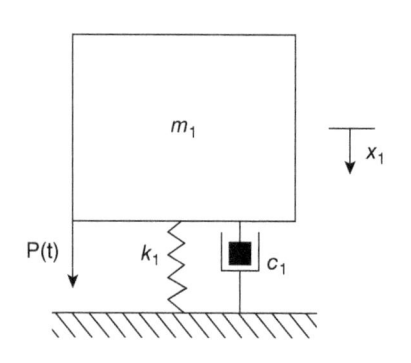

Figure 1.12 Schematic diagram of an idealized system.

those of the primary structure. TMDs are designed to control a single model of a multiple degrees of freedom (MDOF) system (Rana and Soong 1998). Hence, they are tuned to the single structure frequency so that the response in the fundamental mode can be effectively reduced. The addition of a TMD converts a low damping mode in a structure into two coupled higher damping systems with two DOF. The support system for the mass and tuning the frequency are important issues in the design of TMDs. While the mass of the damper is taken as a small fraction of the total mass of the primary structure (usually 1–5%), one of the main limitations is its sensitivity to the narrow frequency band of control. If the TMD is out of tune, its effectiveness is reduced considerably. The primary structure is idealized as a spring-mass SDOF system whose response needs to be controlled (see Figure 1.12). In general, offshore compliant structures exhibit stiff behavior in the vertical plane but remain highly flexible in the horizontal plane. Such behavior offers compliance to the platform without compromising on the payload capacity. The equation of motion of such a SDOF system can be written as:

$$m_1\ddot{x}_1 + c_1\dot{x}_1 + k_1 x_1 = P(t) \tag{1.1}$$

Solving this equation of motion will give the displacement of the mass from its equilibrium position. The natural period of the system is given by the following expression:

$$T_n = 2\pi \sqrt{\frac{m_1}{k_1}} \tag{1.2}$$

When a TMD is attached to the structure, the system becomes a two DOF model, as shown in Figure 1.13. An optimally tuned TMD displaces the damper mass to produce a force on the spring in the direction opposite the forcing function; it controls the response of the primary structure. The TMD absorbs energy from the primary system and forces the structure to move in the opposite direction. The equation of motion of the structure with a TMD can be written as:

$$\begin{bmatrix} m_1 & 0 \\ 0 & m_2 \end{bmatrix} \begin{Bmatrix} \ddot{x}_1 \\ \ddot{x}_2 \end{Bmatrix} + \begin{bmatrix} c_1+c_2 & -c_2 \\ -c_2 & c_2 \end{bmatrix} \begin{Bmatrix} \dot{x}_1 \\ \dot{x}_2 \end{Bmatrix} + \begin{bmatrix} k_1+k_2 & -k_2 \\ -k_2 & k_2 \end{bmatrix} \begin{Bmatrix} x_1 \\ x_2 \end{Bmatrix} = \begin{Bmatrix} P(t) \\ 0 \end{Bmatrix} \tag{1.3}$$

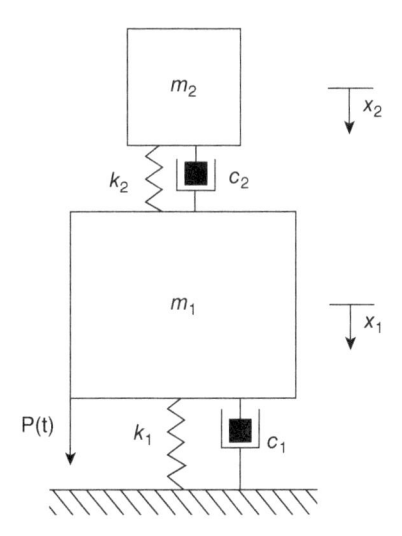

Figure 1.13 Schematic diagram of a spring-mass system with a TMD.

The important parameters that need to be considered in the design of TMDs are the mass ratio and the tuning ratio. While the mass ratio is the ratio of the mass of the damper to that of the primary structure, the tuning ratio is the ratio of the natural frequencies of the damper and the primary structure. The following expressions hold good:

$$T_d = 2\pi \sqrt{\frac{m_2}{k_2}} \qquad (1.4)$$

The natural angular frequency of the damper is ω_2:

$$\omega_2 = \frac{2\pi}{T_d} \qquad (1.5)$$

Mass ratio: $\mu = \dfrac{m_2}{m_1}$ (1.6)

Frequency ratio: $f = \dfrac{\omega_2}{\omega_1}$ (1.7)

Proper tuning of the mass ratio and frequency ratio of the TMD will produce the maximum response reduction of the primary system. The natural frequency of the damper can be varied by changing the mass and the stiffness of the damper. Ahsan Kareem (1983) studied the parameters of human biodynamic sensitivity to building motion. Various means of controlling the motion of high-rise buildings are discussed. A detailed analysis of a dynamic vibration absorber (TMD) for mitigating objectionable levels of motion of tall buildings is studied. An approximate expression is developed to aid in preliminary design procedures. Fujino and Abe (1993) studied the modal properties of TMD structures with a perturbation technique. The mass of the damper is assumed to be small in the analysis. The efficiency of the mass damper is tested for the non-optimal and optimal parameters of the TMD in response to the harmonic, random, free self-excited motion of the structure. Derived formulas help design the TMD for different types of loading. The formulas are based on the tuning ratio, mass ratio, and damping ratio of the primary structure and TMD. Equations for damping a mistuned TMD are also derived. The derived formulas are very accurate for a mass ratio less than 2%.

Rahul Rana and Soong (1998) did a parametric study on the characteristics of TMDs. The performance of TMDs in response to mistuned TMD parameters are analyzed using the time domain and steady-state harmonic excitations. The time-domain analyses are done based on the El Centro and Mexico excitations. Multi-tuned mass dampers (MTMDs) are used to control multiple structural modes. Detuning of the TMD becomes insignificant when the damping of the structure and the mass ratio are high. It is also found that for structures with high damping, the response reduction caused by the TMD is not significant. Chien Liang Lee et al. (2006) presented an optimal design theory in the frequency domain for the response analysis of structures with TMDs. The optimal damping coefficients and stiffness of the system are determined by reducing the structural response performance index. A numerical scheme is presented to identify the optimal design parameters for multiple TMDs and to facilitate convergence effectively and monotonically. The coupled dynamic system of multiple TMDs with the MDOF structure and the power spectral density function of environment loads are considered in the analysis. The optimal design parameters are systematically determined for minimizing the response in the frequency domain.

Wong (2008) investigated the energy transfer process of using a TMD for an inelastic structure to dissipate earthquake forces. The force analogy method is used to model the inelastic structural behavior. A moment-resisting six-story steel frame is considered for the study. Numerical studies are done to understand the energy transfer and the effectiveness of the TMD on the structure under study. Plastic energy spectra are used to find the effectiveness of the damper at different structural yielding levels. The energy stored in the damper is limited when the structure attains the plastic state. If the efficiency of the TMD is reduced, the structural response is the same as if no TMD is installed on the structure. Studies prove that the installation of a TMD increases the energy dissipation of the primary structure by storing energy. For better efficiency, TMDs can be used to extract plastic energy from the lower levels and subsequently release it at the upper level. Tigli (2012) studied the optimum design of dynamic vibration absorbers (DVAs) installed on linear damped systems under random loads. The study is done for the three cases, minimizing the variance of the displacement, velocity, and acceleration of the main mass. A solution for the optimum absorber frequency ratio is obtained as a function of the optimum absorber damping ratio. The mass ratio of the TMD to the modal mass of the building mode responsible for the significant portion of the response is selected to be 0.03 due to practical limitations. From the simulations, it is found that the optimum absorber damping ratio is not significantly related to structural damping. Approximate closed-form optimum design parameters were proposed when the displacement and acceleration variances were minimized. The main advantage of the method is that all the response parameters can be minimized simultaneously.

Viet and Nghi (2014) suggest a nonlinear, single-mass, two-frequency pendulum TMD to reduce horizontal vibration. The proposed TMD contains one mass moving along a bar; the bar can rotate around the fulcrum point attached to the controlled structure. In response to a horizontal excitation, the single TMD mass has two motions (swing and translation) at the same time, and the proposed TMD has two natural frequencies. The nonlinearity of the pendulum is used to increase the number of DOF of the TMD. An approximated solution for the system is provided by solving a scalar algebraic equation. The natural frequencies of the swing and translation motions, respectively, should be tuned to be near the structure frequency and twice the structure frequency.

1.7 Response Control of Offshore Structures

Dong Sheng et al. (2002) experimentally studied the effectiveness of TLDs in reducing the dynamic response of a fixed offshore structure under wave loading. Rectangular and circular TLDs of different shapes and sizes with different water depths are examined. The number of parameters, such as container shape, size, wave characteristics, frequency ratio, and mass ratio, is considered. For maximum reduction of structural response, the frequency ratio should be near to unity. The results showed that three small circular dampers are more effective than a single large circular damper at a low frequency. However, a single large damper is suitable for a higher frequency. As the wave height increases, the reduction in response also increases. When the incident wave period is near twice the fundamental structure period, the maximum reduction is obtained. The response of the structure can be reduced for a wide range of frequencies. It is suggested that this type of damper could be used for fixed offshore structures in the presence of random waves.

Lee et al. (2006) studied the response reduction of a floating platform attached to a TLCD given wave-induced vibrations. The stochastic analytical method in the frequency domain is utilized, and due to this, the linearization scheme for the system is applied. The Morison equation for a small body is used to calculate the wave forces. Parametric studies are done for the draft and size of the pontoons. A model test is conducted to check the feasibility of the TLCD device applied to the platform. It is found that a perfectly tuned TLCD reduces the response at the resonant frequency. Analytical results show that the energy dissipated from the TLCD device on the floating platform system may be from 50 to 70%. The draft and dimensions of the platform influence the performance of the TLCD. The mass variation does not have much effect on the performance of the TLCD. Qiao Jin et al. (2007) studied the efficiency of circular TLDs on reducing the seismic response of a jacket structure with a liquid sloshing experiment, a model experiment, and numerical analysis. A numerical investigation is carried out using the lumped mass method. A higher mass ratio of the TLD is very efficient for reducing the response of the platform under earthquake loading. TLDs are found to be effective for a mass ratio from 0.01 to 0.05.

Taflanidis (2008, 2009) proposed a robust stochastic design approach for appropriate tuning of the TMD. A TLP having a different natural period for the pitch and heave motions is considered for the study given random sea conditions. The application of two mass dampers is suggested for efficient reduction of pitch and heave motions. TMDs modeled as secondary masses are connected to the hull of the TLP with the help of a dashpot and springs. The study is done to understand the effect of a single damper and two dampers attached to the hull to increase the safety of the platform. The results showed that the use of two dampers in parallel operation was effective in the response control of heave and pitch motions. Colwell and Basu (2009) did simulation studies on the structural response of an offshore wind turbine attached to a TLCD. Environmental loading is considered by combining the wind loads from the Kaimal spectrum and the wave loads from the JONSWAP spectrum. Numerical simulation is carried out, modeling the turbine tower as a MDOF system. Blades at the stationary position and rotating condition were investigated. The system with a TLCD showed reduced bending moment and displacement. The peak response is reduced by 55%. Also, the fatigue life of the wind turbine is enhanced.

Chandrasekaran et al. (2010) studied dynamic response characteristics such as bending stress variations and the displacement of a multi-legged articulated tower (MLAT) through experiments. A TMD is attached to the bottom of the deck plate. Its natural frequency is tuned near to the natural frequency of the MLAT vibration mode that is to be controlled. The MLAT itself is modeled as a SDOF mass-spring-damper system. A significant reduction in the bending moment of the tower is observed for higher wave heights. The maximum reduction is observed at a near-resonant frequency of the structure. Lee and Juang (2012) proposed a new concept of an underwater tuned liquid column damper system (UWTLCD). The study focused on increasing structural integrity by reducing the response of the structure to the incident wave and stresses on the structure. A TLCD with smaller horizontal tubes is pooled into the pontoon of the TLP. Experimental studies were done to find the effectiveness of the UWTLCD in reducing the response. A parametric study was conducted on the effect of wave conditions, the height of the pontoon, and the liquid-length of the TLCD. The results prove that a properly tuned UWTLCD system is very effective at reducing the hydrodynamic response and the tensile force on the tethers. It is shown that a UWTLCD can effectively reduce the heave response.

Moharrami and Tootkaboni (2014) proposed an innovative concept for reducing the displacement response of a tower fixed offshore platform to wave loads. A hydrodynamic buoyant mass damper (HBMD) uses the damper's buoyancy and inertial forces along with hydrodynamic damping effects to reduce the displacement response of the platform. A jacket fixed platform is investigated in which the HBMD is added at the appropriate elevation. As a result of the damper's eccentricity for the platform's position in its deformed configuration, the upward buoyancy force causes a reversal moment that can potentially counteract forces generated by wave loads. The HBMD moves through a surrounding fluid and creates reversal forces on the structure, which moderate those from the wave loads. The reversal forces consist mainly of the inertial force and the buoyancy force. In addition to inertia and buoyancy forces, forces associated with eddy formations in the proximity of the HBMD also help to reduce the platform's response under wave loading.

1.8 Response Control of TLPs Using TMDs: Experimental Investigations

Experimental studies on response control of a TLP in the presence of a TMD are discussed (Ranjani 2015; Chandrasekaran et al. 2016). A scale TLP model is fabricated at a 1 : 100 scale ratio using Froude scaling to include the effect of gravity forces and wave resistance in the model. Experiments are carried out for three different cases: (i) TLP model without a TMD (Case 1), (ii) TLP model with a TMD of mass ratio 1.5% attached to the structure (Case 2), and (iii) TLP model with a TMD of mass ratio 3.0% attached to the structure (Case 3). They are examined in the presence of both regular and random waves. The geometric sizing of members and the plan dimensions of the assumed model are derived from the Auger TLP, Gulf of Mexico. Acrylic material is used to fabricate four columns of 250 mm outer diameter and 5 mm thickness; the height of the column is 490 mm. The bottom and top of the column members are closed with an acrylic sheet of 10 mm thickness. An additional sheet of 10 mm thickness is placed inside each column to provide the required lateral stiffness for the members. The spacing between the longitudinal columns is 850 mm, while the spacing between the transverse columns is 650 mm. Pontoons of the rectangular cross-section of size 90 × 110 mm are used in the model. The deck of the TLP is made of a 1000 mm square acrylic sheet of 5 mm thickness. Clearance between the column and deck is fixed at 25 mm, and a draft of 300 mm is maintained.

The TMD considered in this experiment consists of a solid mass and a spring element. The mass is fabricated from 10 mm thick acrylic sheets. A rectangular 120 mm × 100 mm × 850 mm box is fabricated, as shown in Figure 1.14. The mass of the rectangular box is 0.6 kg ($\mu = 1.5\%$). A mass ratio of 3% is achieved by adding 0.6 kg of sand to the box. To ensure free movement of the attached secondary mass, frictionless rollers 40 mm in diameter are attached to the base plate of the box with a clearance of 15 mm. Two sides of the box are fixed with a ring arrangement for attaching the spring, as shown in the figure. The spring is made of steel wire 1 mm in diameter. A hook is formed at both ends of the spring to connect the mass. The spring is fabricated with a constant outer diameter of 50 mm for both models, as shown in the figure. Several active coils are chosen with 30 and 15

Figure 1.14 Mass used in the TMD.

Table 1.3 Properties of a TMD in the model and prototype (scale 1 : 100).

	Model		Prototype	
Description	Case 2	Case 3	Case 2	Case 3
Mass of damper	0.6 kg	1.2 kg	600 T	1200 T
Outer diameter of spring	50 mm	50 mm	5 m	5 m
No. of active coils	30	15	30	15
Pitch (mm)	0.5 mm	0.5 mm	0.05 m	0.05 m
Stiffness of spring	2.76 N/m	5.87 N/m	27.6 kN/m	58.7 kN/m

turns to suit the required mass ratios of 1.5% and 3.0%, respectively. The model details of the TMD are given in Table 1.3.

The model tests are conducted in the presence of regular waves with a wave height of 4, 8, and 12 cm, at a wave heading angle of 0° in a period ranging from 1.2 to 4.4 seconds with a time interval of 0.2 seconds. There was sufficient waiting time between each test to achieve calm water conditions. Experimental investigations are also carried out with random waves of significant wave height (8 cm) and a period ranging from 1.2 to 3.25 seconds. The PM spectrum is used to obtain random sea conditions, given by the following expression:

$$S(\omega) = 5\pi^4 \frac{H_s^2}{T_p^4} \frac{1}{\omega^5} exp\left(-\frac{20\pi^4}{T_p^4} \cdot \frac{1}{\omega^4}\right) \tag{1.8}$$

Table 1.4 Results of free oscillation tests of the TLP model.

	TLP surge		Pitch		TMD surge		Difference %	
	T_n (s)	ζ (%)	T_n (s)	ζ (%)	T_n (s)	ζ (%)	T_n (s)	ζ (%)
TLP without damper	4.02	9.97	0.5	35.69	—	—	—	—
TLP with damper ($\mu = 1.5\%$)	4.2	13.65	0.46	35.98	2.2	8.98	4.47	36.9
TLP with damper ($\mu = 3.0\%$)	4.4	16.69	0.45	36.25	2.2	12.10	9.45	67.4

The time history of the wave height, surge, heave responses, and variation in tether tension are recorded simultaneously for a period of about 60 seconds. Table 1.4 shows the results of free-oscillation tests, with and without dampers.

It can be seen that the presence of the damper increases the surge period and damping ratio for the chosen mass ratios. When the TMD is attached to the TLP, the mass of the TLP increases for the chosen mass ratio. It results in an increase in the period of the platform in the presence of the TMD. The surge period increases up to 4.47 and 9.45% for the two different mass ratios under consideration. It is observed that the damping ratio of the structure increases by 36.9 and 67.4% in the surge. However, the presence of the TMD does not significantly influence the natural period and the damping ratio of the structure in the stiff DOF (for example, pitch). The inertial force generated by the TMD controls the response of the platform. Maximum response control is expected when the responses of the platform and TMD are out of phase with each other. Figures 1.15 and 1.16 show the surge-displacement time history of both the platform and the TMD for two different mass ratios with a wave height of 12 cm and a period of 1.2 seconds. It is seen that the response of the TMD is out of phase with the structure response, which is necessary to achieve effective control.

It is known that the response of the TMD will increase with the increase in the period of the damper, which may result in a larger displacement of the TMD beyond the deck space. Figure 1.17 shows the surge RAO of the TMD for different mass ratios. At lower wave periods, the response of the TLP is much less, and the TMD does not get enough energy to become active. Hence the response of the TMD is also much less. As the wave period increases, the response of the TLP increases, which excites the TMD to come into operation. The response of the TMD increases until it reaches its natural period. Since the response of the TMD is restricted, it was observed that from the natural period of the TMD, the TMD had a response until the maximum possible limit. Hence a constant response is obtained.

Response control of a TLP with a TMD is also examined in the presence of random waves. A comparison of the surge acceleration spectrum for a TLP with and

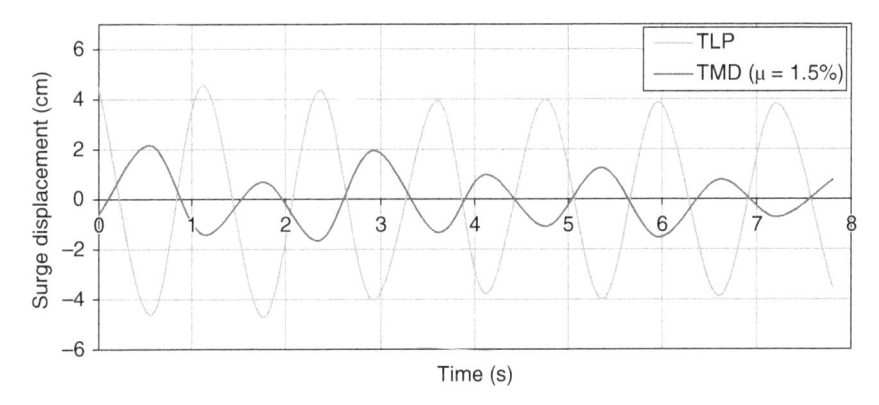

Figure 1.15 Response of a TLP and TMD (μ = 1.5%).

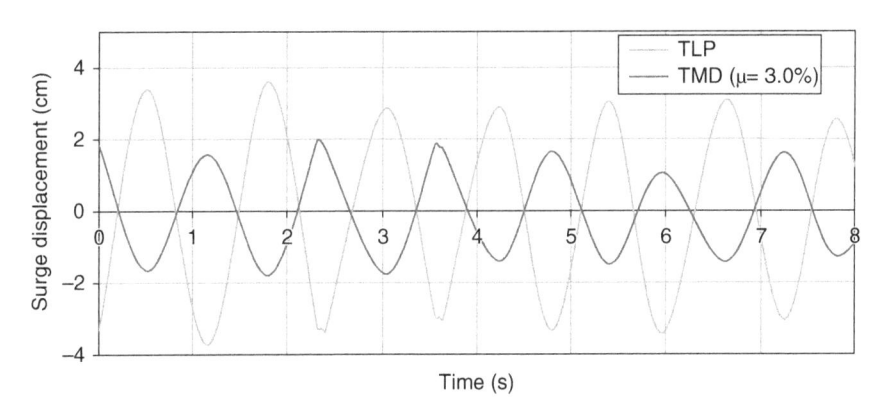

Figure 1.16 Response of a TLP and TMD (μ = 3.0%).

without a damper for a significant wave height of 8 m at various peak periods is shown in Figures 1.18–1.21. The peak surge response is observed near the peak period of the excitation wave. It is also seen that the addition of a TMD reduces the peak amplitude of the surge response considerably. The root mean square (RMS) values of the surge response of the TLP with and without a damper in response to various random excitations are summarized in Table 1.5.

A comparison of pitch response with and without a damper for a significant wave height of 8 m at various periods is shown in Figures 1.22–1.25. It is seen from the figures that the peak response is observed near the peak period of the excitation wave, while the presence of a TMD causes a significant reduction in peak amplitude of the response. The increase in mass ratio reduces not only the peak amplitude but also the overall response of the TLP. The RMS values of the pitch

42 | *Offshore Compliant Platforms*

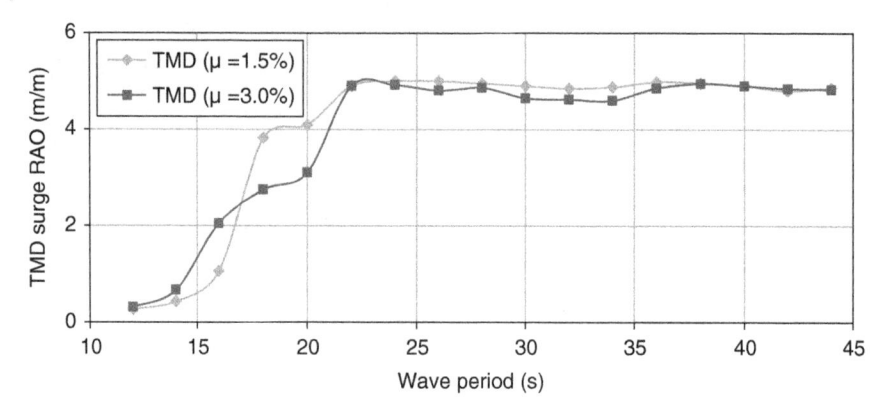

Figure 1.17 Surge RAO of a TMD.

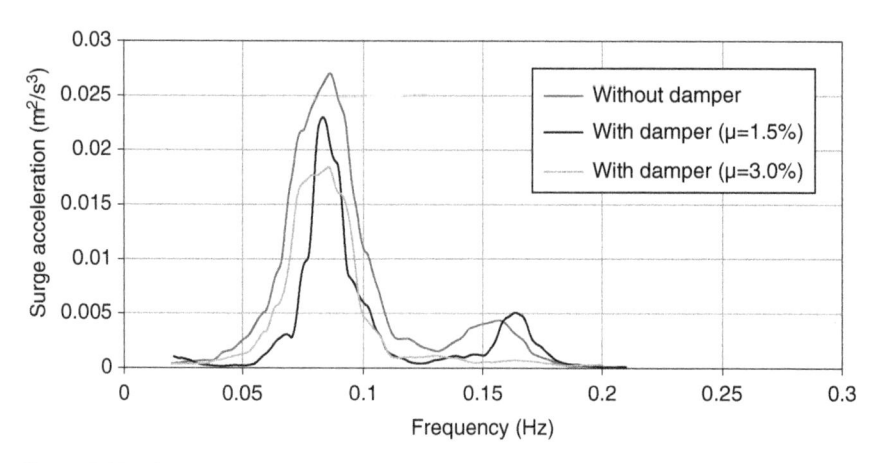

Figure 1.18 Comparison of surge response (H_S = 8 m; T_P = 12 seconds).

response with and without a damper in response to various random excitations are summarized in Table 1.6.

The TLP produces a higher response at longer wave periods and increased wave heights. This response needs to be controlled, which can be achieved by using a TMD with a higher mass ratio. It is also evident that considerable energy is available with higher amplitude waves to activate the TMD motion. Experimental investigations carried out on the scale model of a TLP in the presence of a TMD confirmed the following:

i) A spring-mass system with a higher mass ratio is effective for response reduction with a wide range of periods.
ii) A TMD shows better control for larger wave heights.

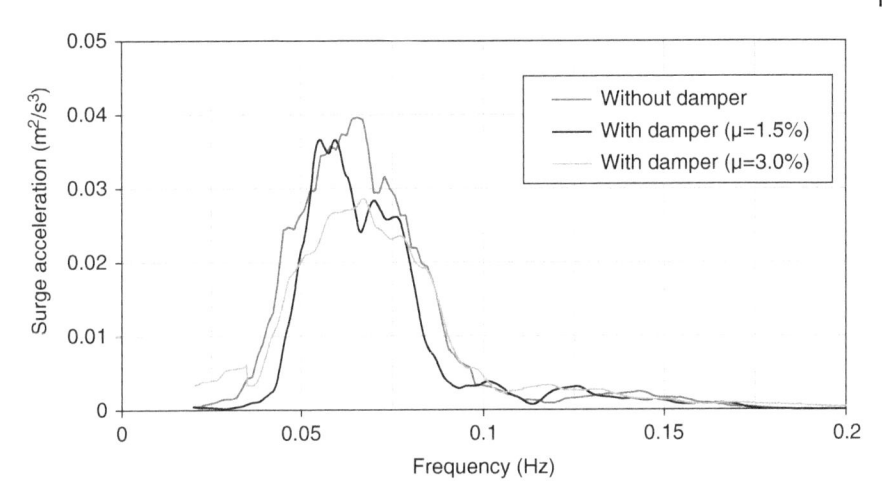

Figure 1.19 Comparison of surge response (H_S = 8 m; T_P = 16 seconds).

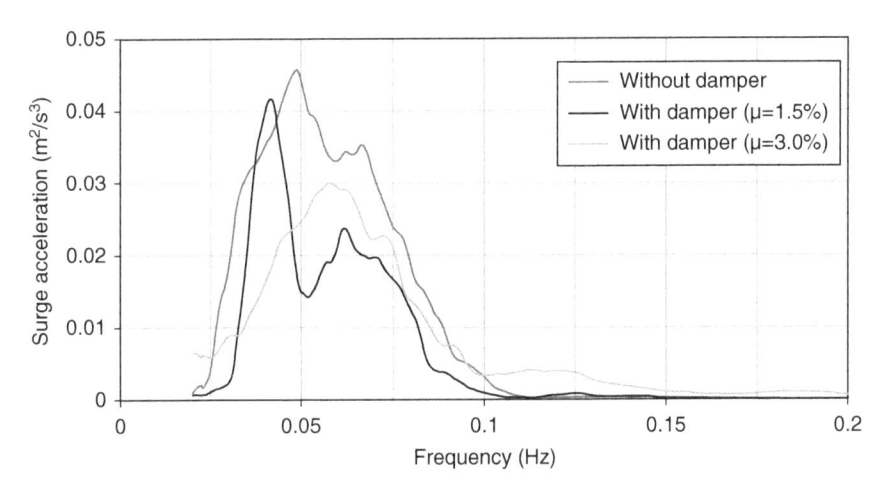

Figure 1.20 Comparison of surge response (H_S = 8 m; T_P = 20 seconds).

iii) An increase in wave elevation increases the surge response at higher periods.
iv) Adding a TMD to the structure shifts the surge, heave, and pitch natural periods and increases the damping ratio of the structure.
v) The response reduction is found to be high for higher mass ratios. Maximum response reduction, up to 10.9 and 16%, is obtained mass ratios of 1.5 and 3.0%, respectively.
vi) Greater heave response reduction is observed due to the reduction of the surge response and tether tension variation. A maximum reduction of 19 and 28% is obtained in the heave response for mass ratios of 1.5 and 3.0%, respectively.

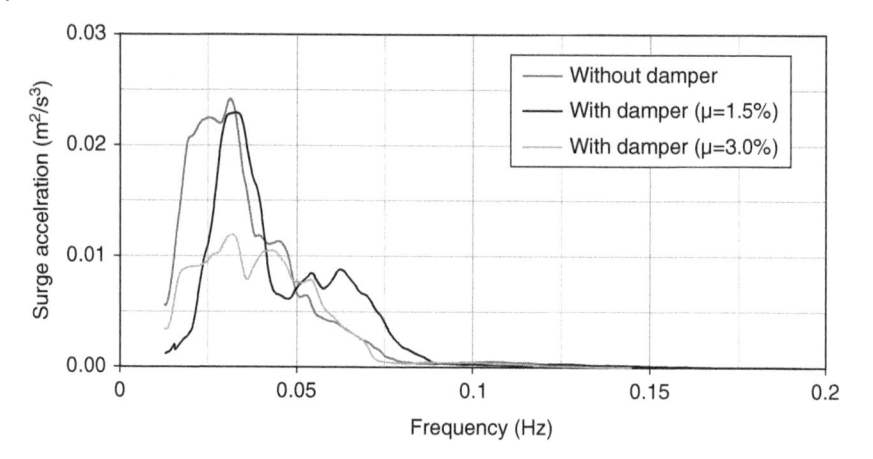

Figure 1.21 Comparison of surge response (H_S = 8 m; T_P = 32.5 seconds).

Table 1.5 RMS value of surge responses in the presence of random waves.

Period	TLP without damper	TLP with damper (1.5%)	TLP with damper (3.0%)	Response reduction % (μ = 1.5%)	Response reduction % (μ = 3.0%)
12	0.104	0.095	0.087	8.6	16.6
16	0.126	0.117	0.103	7.5	18.5
20	0.140	0.130	0.119	7.2	15.1
32.5	0.155	0.146	0.135	5.9	12.9

vii) By controlling the surge response, indirect control of the heave and pitch DOF is achieved. A maximum reduction of 13 and 16% is obtained in pitch for mass ratios of 1.5 and 3.0%, respectively.

1.9 Articulated Towers

Articulated towers are semi-compliant offshore structures consisting of a universal joint that connects the tower to the seabed. The presence of a universal joint allows the structure to move and thereby reduces the forces acting on the tower. The articulated tower (AT) has emerged as one of the most reliable systems for single-point mooring, control towers, and flare structures but can also be used as a production platform for marginal fields, or as a processing unit in remote, hostile environments. This kind of platform can be seen as an extension of the TLP in

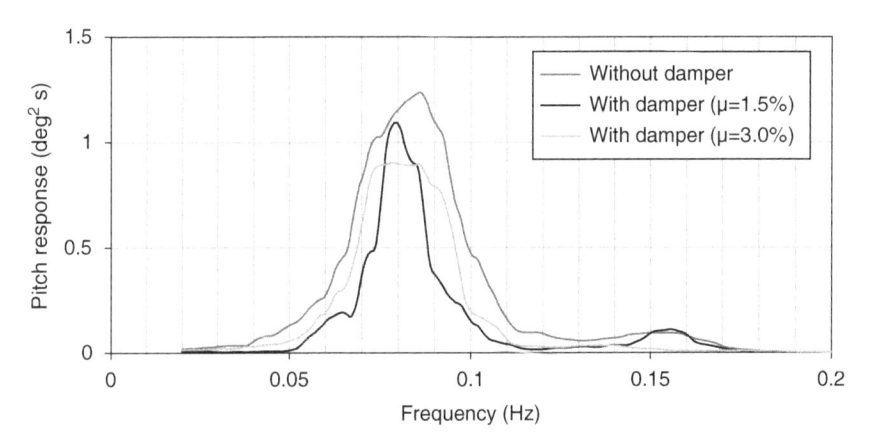

Figure 1.22 Comparison of pitch response (H_S = 8 m; T_P = 12 seconds).

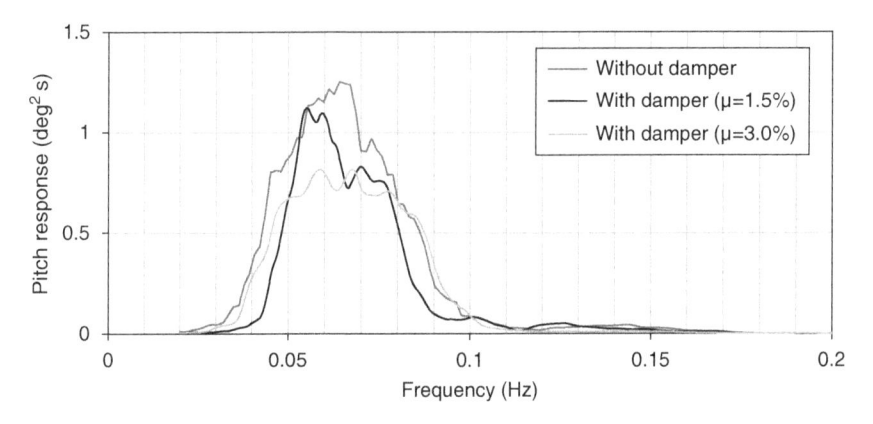

Figure 1.23 Comparison of pitch response (H_S = 8 m; T_P = 16 seconds).

which the tension cables are replaced by one single buoyant shell with sufficient buoyancy to produce required restoring moment against lateral loads. An articulated tower is then flexibly connected to the seabed through a universal joint and held vertically by the buoyancy force acting on it. Similar to a reed that "bends but does not break," the suppleness of articulated structures withstands the combined effects of wind and waves. As the connection to the seabed is through the articulation, the structure is free to oscillate in all directions and does not transfer any bending moment to the base.

The first AT was built and operated in the Argyll Field in the North Sea in 1975. The basic configuration of an AT is shown in Figure 1.26. It comprises five cylindrical subsections erected consecutively in the vertical plane: the connector at the lower part, ballast chamber, lower shaft, buoyancy chamber, and upper shaft. The

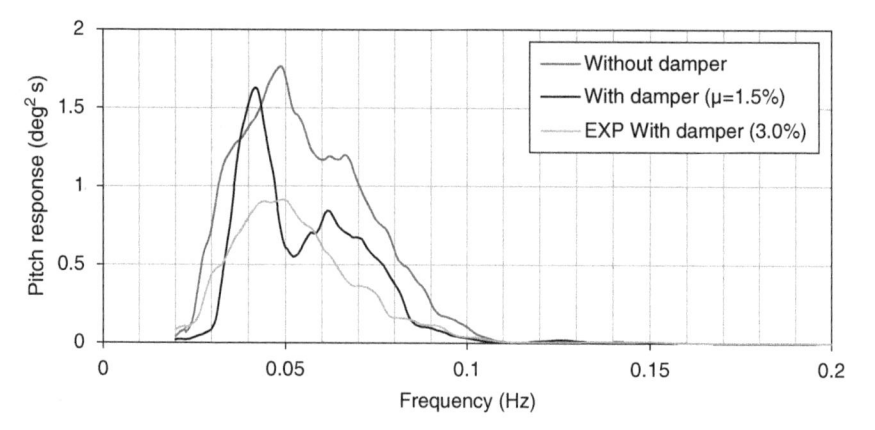

Figure 1.24 Comparison of pitch response (H_S = 8 m; T_P = 20 seconds).

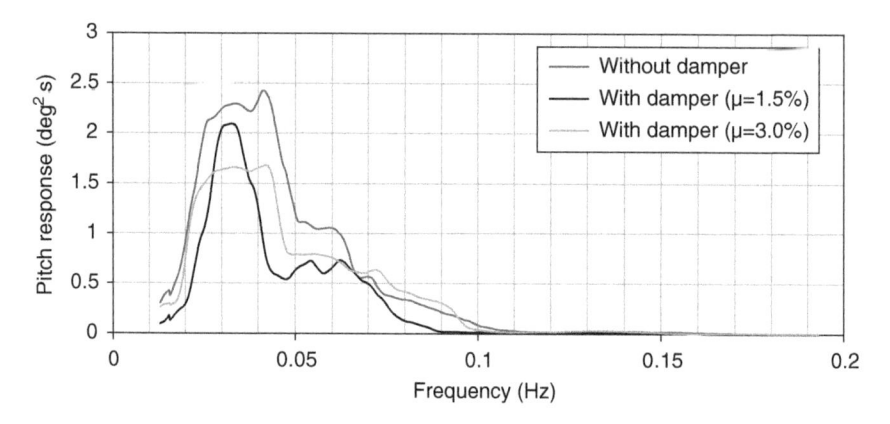

Figure 1.25 Comparison of pitch response (H_S = 8 m; T_P = 32.5 seconds).

connector is joined to the base at the sea bottom and called a *universal joint*, while the upper shaft supports a deck structure where necessary topside facilities are accommodated. Unlike fixed structures, which are designed to withstand environmental forces without substantial displacement, ATs are designed to allow small but not negligible deformation and deflection that are made possible by the presence of the universal joint. The utilization of the universal joint also relieves the foundation from resisting any lateral force developed by environmental action.

One of the primary features of an AT is its ability to displace from its initial position when subjected to environmental loads, hence reducing the maximum internal response of its structural elements. Under environmental loads, the AT displaces in the rotational mode of the universal joint located on the base. A large buoyancy chamber enables recentering of the tower, once displaced. The buoyancy chamber

Table 1.6 RMS value of pitch responses in the presence of random waves.

Period	TLP without damper	TLP with damper (1.5%)	TLP with damper (3.0%)	Response reduction % ($\mu = 1.5\%$)	Response reduction % ($\mu = 3.0\%$)
12	0.683	0.604	0.546	11.5	20.0
16	0.695	0.597	0.537	14.1	22.7
20	0.840	0.755	0.690	10.2	17.9
32.5	0.962	0.848	0.776	11.8	19.4

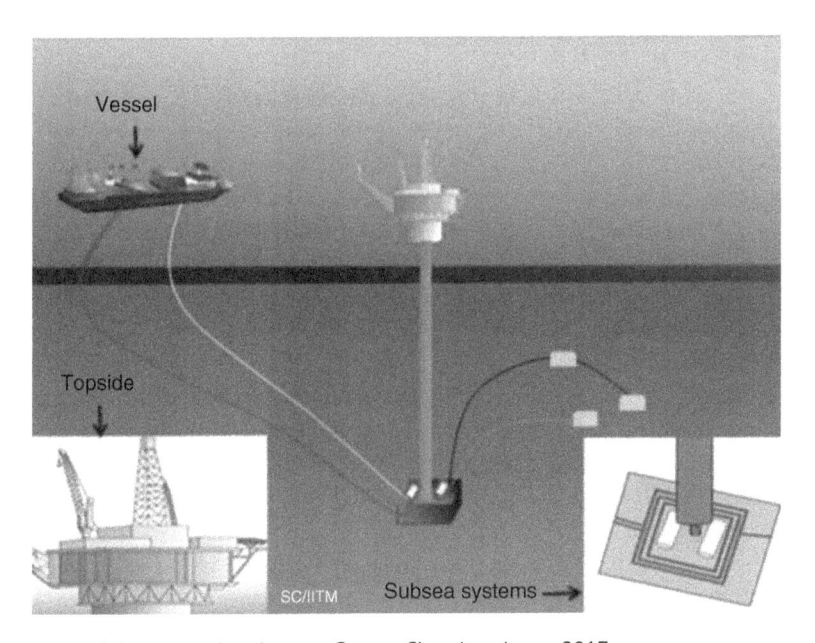

Figure 1.26 Articulated tower. *Source:* Chandrasekaran 2017.

is considered the most important element, since it provides essential stiffness to the system through buoyancy-restoring forces. In a typical AT, when a moored tanker pulls the top of the tower to one side, tilting the tower, the buoyancy compartment moves as an arc about the pivot pint to recenter the tower. However, in shallow water, a small angle of tilt of the tower results in a decreasing moment arm, which proportionately decreases the restoration force. It can be compensated for by utilizing a larger buoyancy compartment, but this results in a large, costly system. The use of a buoyancy chamber in place of guy lines or a tether is required to restrain any possible excessive motions, thus simplifying the system even further.

ATs are considered economically attractive in deepwater applications due to their reduced structural weight and simplicity of fabrication, compared to other conventional platforms. They are well suited for water depths ranging from 150 to 500 m. One advantage of this type of system over those that utilize a freely floating buoy held by catenary chains is the fact that fluid-carrying conduits can be placed to extend through the rigid tower to protect these conduits. Another advantage is that the tower can extend high above the water to provide an attachment point for a hawser connected to a ship, without destabilizing the system. As the water depth increases, it is preferable to add another universal joint along the height of the tower. An AT with universal joints in the intermediate level is called a single-leg multi-hinged AT. The extension of the concept of the single-leg AT led to the development of a new type of platform with several columns that are parallel to one another: the MLAT.

A MLAT is an AT in which, instead of a single shell, the deck is attached to the seabed by three or four legs that are parallel to one another and connected by universal joints both to the deck and to the foundation. The use of universal joints ensures that the columns always remain parallel to one another and the deck remains in a horizontal position. There is no rotation about the vertical axis of the columns. The advantage of this system is that the payloads and deck areas can be increased. Further, they are comparable to conventional production platforms in moderate water depths; and the sway or horizontal displacement of the deck is considerably reduced compared to single-leg ATs, making the use of such platforms feasible even as production units in deep water or ultra-deep water. In this sense, it is important to reduce the displacement of the superstructure as much as possible. This can then enable the installation of more facilities that are needed for production, and also ensure save living units on the topside. One of the drawbacks is that while for a single- or multi-hinged AT, the high period generally avoids resonance problems, these structures are characterized by closer periods of waves (occurring every 10–20 seconds) and could be subjected to resonance problems. Under such conditions, the platforms' safe operability needs to be guaranteed through effective design; but the literature is still poor, offering no feasible solution until now.

1.10 Response Control of ATs: Analytical Studies

A system on which a steady alternating force of constant frequency is acting may result in undesired vibrations, especially when the excitation frequency is close to the structure's fundamental frequency. In order to solve this problem, one may try to eliminate the force bandwidth. One can also attempt to change the structural characteristics of the system. However, in the case of offshore structures, the latter option influences the buoyancy of the system. An alternative solution to reduce the undesired vibration response is to apply a kind of DVA to the system, such as a TMD. The TMD chosen for the current application is a pendulum, as this is a simple and efficient tool to control the response. The MLAT is modeled as a SDOF mass-spring-damper system on which is attached the TMD, which consists of a comparatively small vibratory system k_2, m_2, attached to the main mass m_1. Such a system is represented schematically in Figure 1.27.

The equation of motion governing this system is given by:

$$\begin{cases} m_1\ddot{x}_1 + c_1\dot{x}_1 + k_1 x_1 + c_2(\dot{x}_1 - \dot{x}_2) + k_2(x_1 - x_2) = P_0 \sin\omega t \\ m_2\ddot{x}_2 + c_2(\dot{x}_2 - \dot{x}_1) + k_2(x_2 - x_1) = 0 \end{cases} \quad (1.9)$$

in which both x_1 and x_2 are harmonic motions of the frequency (ω) and can be represented as vectors. The easiest manner of solving this system is to write these vectors as complex numbers. The steady-state solution is represented by the following equation:

$$\begin{cases} x_1 e^{j\omega t} \\ x_2 e^{j\omega t} \end{cases} \quad (1.10)$$

Figure 1.27 Analytical model.

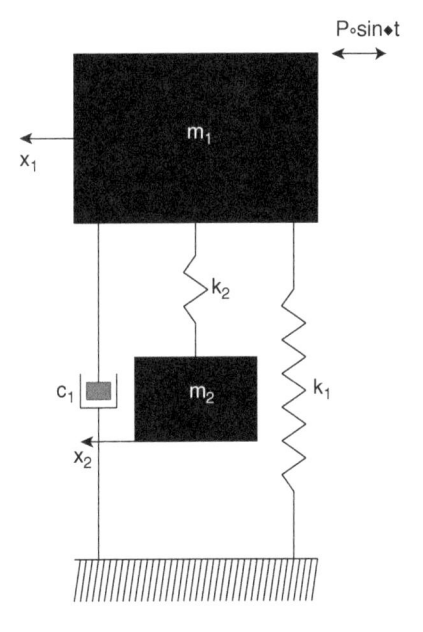

where x_1 and x_2 are complex numbers. Differentiating these solutions, substituting them in Eq. (1.9), and dividing for $e^{j\omega t}$ transforms the differential into the algebraic equation show here:

$$\begin{cases}(-m_1\omega^2+c_1 j\omega+k_1+k_2)x_1-k_2 x_2 = P_0 \\ (-m_2\omega^2+k_2)x_2-k_2 x_1 = 0\end{cases} \quad (1.11)$$

Solving for x_1, we get:

$$\left(\frac{x_1}{P_0}\right)^2 = \frac{(-m_2\omega^2+k_2)^2}{\left[\left(-m_1\omega^2+k_1\right)\left(-m_2\omega^2+k_2\right)-k_2 m_2\omega^2\right]^2+c_1^2\omega^2\left(-m_2\omega^2+k_2\right)^2} \quad (1.12)$$

from which one can see that x_1 is a function of seven variables: P_0, ω, c_1, k_1, k_2, m_1, and m_2. For simplification, we can put this equation in a dimensionless form by introducing the following symbols:

$x_{st} = P_0 / k_1$

$\omega_a^2 = k_2 / m_2$

$\Omega_n^2 = k_1 / m_2$

$\mu = m_2 / m_1$

$f = \omega_a / \Omega_n$

$g = \omega / \Omega_n$

$c_c = 2m\omega_a$

where x_{st} is the static deflection of the main system, ω_a^2 is the natural frequency of the absorber, Ω_n^2 is the natural frequency of the main system, μ is the mass ratio, f is the natural frequency ratio, g is the forced frequency ratio, and c_c is the critical damping. After simplifications, Eq. (1.12) reduces to the following simple form:

$$\frac{x_1}{x_{st}} = \sqrt{\frac{(-f^2+g^2)^2}{\left[\left(-g^2+1\right)\left(-f^2+g^2\right)-\mu f^2 g^2\right]^2+4\frac{c_1^2}{c_c^2}g^2 f^2\left(-\mu f^2+\mu g^2\right)^2}} \quad (1.13)$$

It can be seen that this equation expresses the amplitude ratio in terms of four essential variables instead of seven: μ, c_c, f_2, and g_2. This simplified form of the response of the primary system exhibits a variation, as shown in Figure 1.28 for a

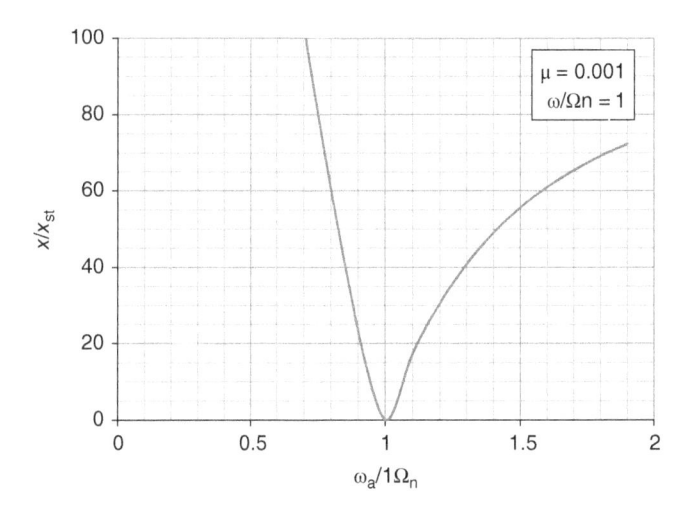

[Figure: plot with x-axis $\omega_a/1\Omega_n$ from 0 to 2, y-axis x/x_{st} from 0 to 100, showing a sharp V-shape minimum near 1. Legend: $\mu = 0.001$, $\omega/\Omega n = 1$.]

Figure 1.28 Variation of responses for different frequency ratios.

mass ratio of 0.1% and a frequency ratio of unity. We can see that the range of the absorber frequency is narrow for an effective reduction of displacement.

However, by considering a secondary mass as a TMD, the variation of response of the primary structure with a TMD of the same frequency based on the action of waves can be controlled. Figure 1.29 shows the variation of response for different frequency ratios with a TMD. It is seen that such a system can avoid displacement at the resonance frequency. This is due to the fact that the inertial force of the TMD, at all instants, acts opposite in phase with the excitation force $P_o \sin \omega t$. It results is no force acting on the main system and hence does not vibrate the primary system at all. Moreover, this also shows that the presence of a TMD can reduce displacement even when the excitation frequency is smaller than the structural frequency. Hence, for structural systems subjected to a variable frequency excitation, as in the present case, the addition of a TMD can result in additional problems because it generates two resonant peaks instead of one. However, interestingly, response build-up, even given these near-resonance conditions, can be controlled by introducing a damper in the TMD (Den Hartog 1985).

The influence of the mass ratio in the system is shown in Figure 1.30. It is evident that the increase in mass ratio increases the response reduction significantly for the same frequency ratio of unity. Unfortunately, this is not a feasible solution for offshore structures as an increase in mass affects buoyancy and thereby stability. Hence, beyond a threshold value, it is not possible to control the response using a TMD. In the following section, experimental investigations carried out on response reduction of an AT using a TMD are discussed.

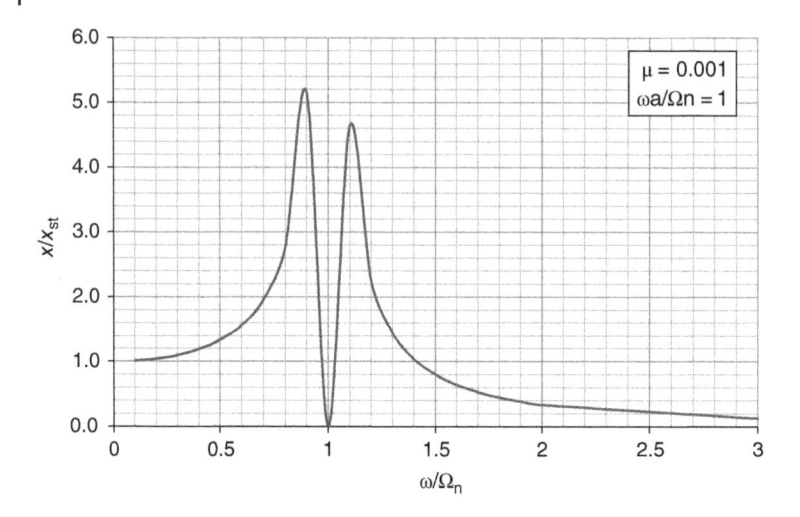

Figure 1.29 Variation of responses for different frequency ratios with a TMD.

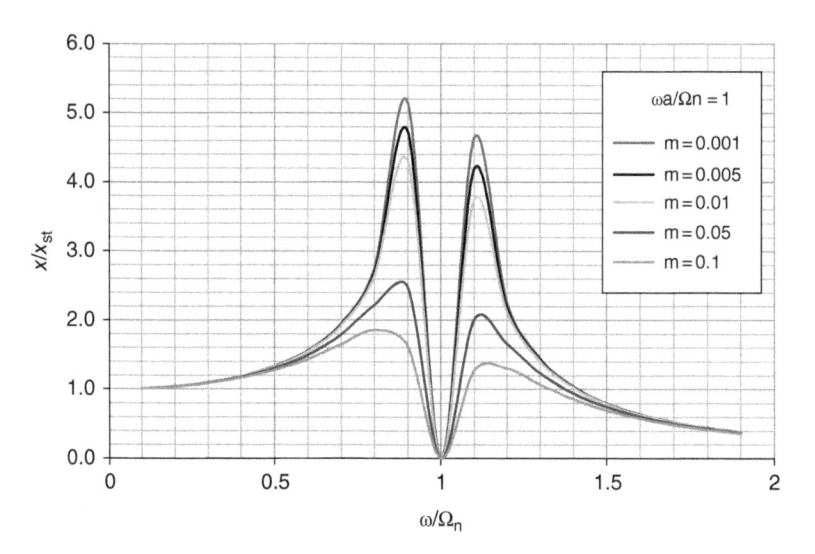

Figure 1.30 Variation of responses for different frequency and mass ratios with a TMD.

1.11 Response Control of ATs: Experimental Studies

A scale model of a MLAT is fabricated to a scale of 1 : 100. Perspex is used to fabricate the model and ensure that during test conditions, stresses that develop in the members do not exceed the elastic limit of the chosen material. The mechanical characteristics of Perspex are summarized in Table 1.7.

Table 1.7 Mechanical properties of the Perspex material used for the model.

Properties	Units	Value
Design modulus (Up to 25°)	Gpa	17
Design stress (Up to 25°)	Mpa	2.5
Density	Kg/m^3	1.19
Coefficient of thermal expansion	K^{-1}	
Poisson's ratio		0.38

Universal joints are used to connect the legs of the tower to the base, while the other end of the leg is connected to the deck. All four legs are fabricated from eight tubes 60 cm long, coupled by an internal ring with a 9 cm diameter and joined with chloroform to make the connection waterproof. Both the top and bottom ends of the tubular legs are covered with two Perspex plates each and made watertight. Subsequently, the four legs are connected to a steel plate at the bottom. One of the legs is furnished with five strain gauges of type FLA-3-350-11 from Tokyo Sokkei Kenkyujo (Japan) at a spacing of every 20 cm, in order to evaluate the strain along the leg. After fixing the strain gauges, waterproofing paste is applied over the strain gauges. Figure 1.31 shows the view of the model with details. Figure 1.32 shows the model of the TMD.

The TMD is chosen to fit the close-resonance band, whose period is given by the following relationship:

$$T = 2\pi\sqrt{\frac{l}{g}} \tag{1.14}$$

Legs of three different lengths are chosen: 12, 19, and 27 cm, which correspond to the periods 0.74, 0.88, and 1.05, respectively. The periods of the TMD with chosen lengths are experimentally evaluated using a potentiometer and an oscilloscope to ensure the design values.

1.11.1 MLAT Without a TMD

Experimental investigations are first carried out on the scale TLP model under wave loads without a TMD. Figure 1.33 shows the free-vibration time history of the tower, with which the natural period of the tower is estimated to be 2.69 seconds; this is also confirmed by the numerical model using the software. Subsequently, the MLAT model is subjected to 3 cycles of waves, each made of 12

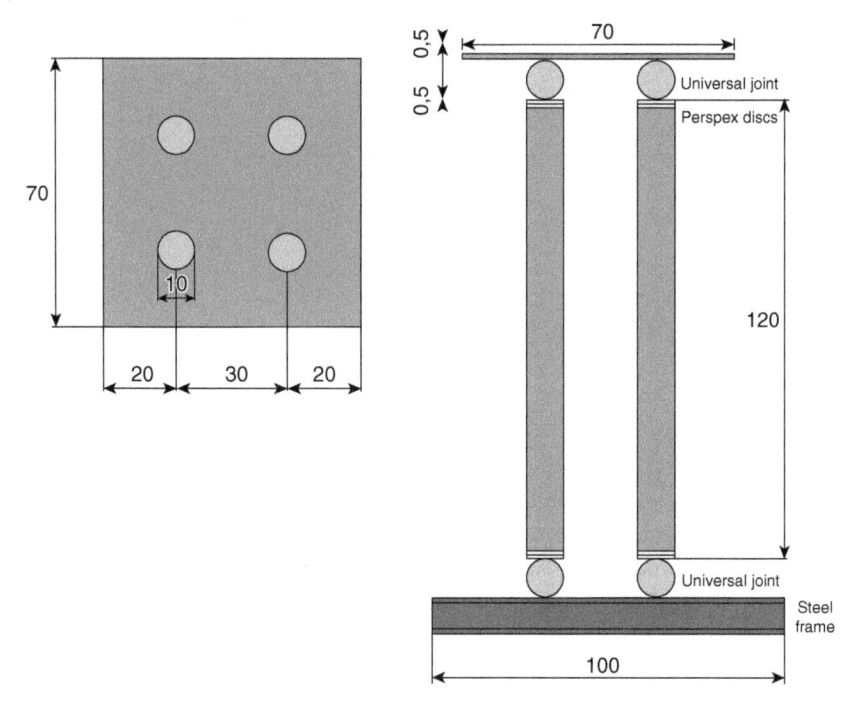

Figure 1.31 Geometric details of the model.

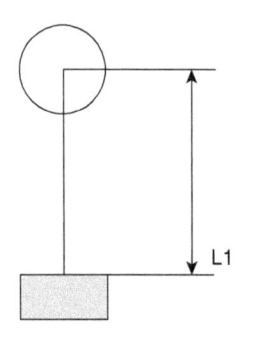

Figure 1.32 Model of a TMD.

waves of different periods: 0.7, 0.8, 0.9, 1.0, 1.1, 2.0, 2.2, 2.4, 2.6, 2.7, 2.8, and 2.9 seconds. Figure 1.34 shows the surge RAO of the tower without a TMD when subjected to three different wave heights. It is seen that surge displacement increases almost linearly up to the first peak, at 2.6 seconds. A kink in the response for all wave heights is attributed to slip that occurred in the tower response, which is characterized by large response amplitudes due to the presence of universal joints. The response further increases at 2.8 seconds with the increase in wave

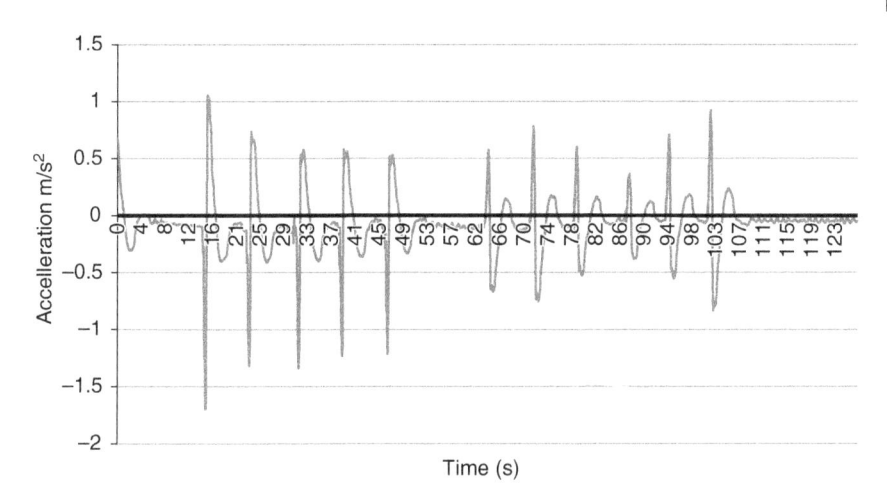

Figure 1.33 Free-vibration time history.

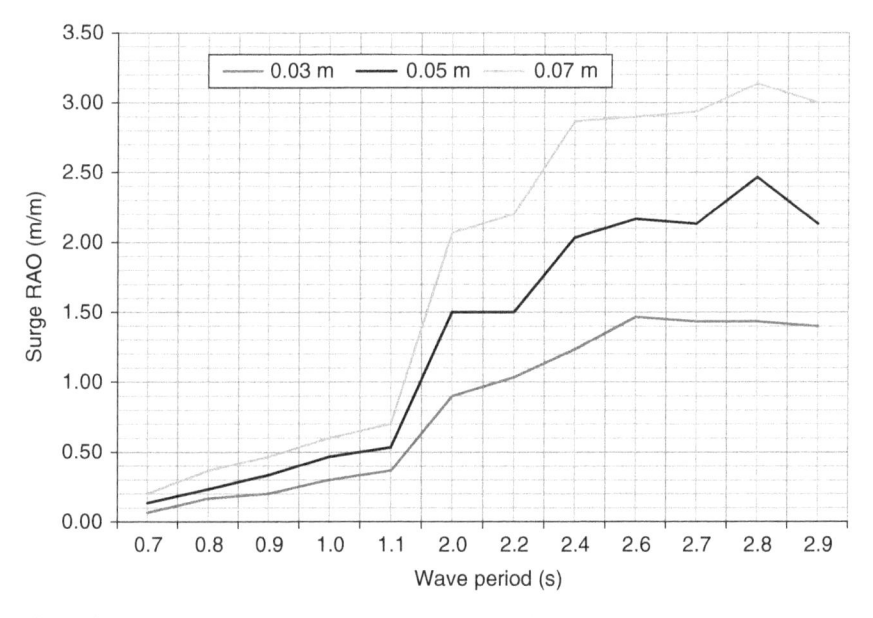

Figure 1.34 Surge response of a MLAT without a TMD.

height. This is explicable because even if the structure is modeled with a single DOF, in the real application, the platform exhibits a MDOF characteristic, showing more than one vibration mode. Hence, the second peak corresponds to the frequency of the structure's second vibration mode.

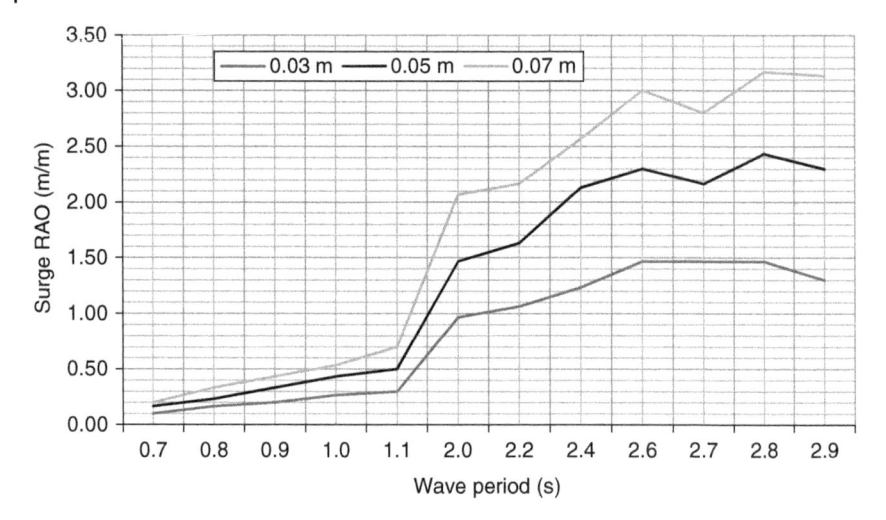

Figure 1.35 Surge RAO for TMD-1.

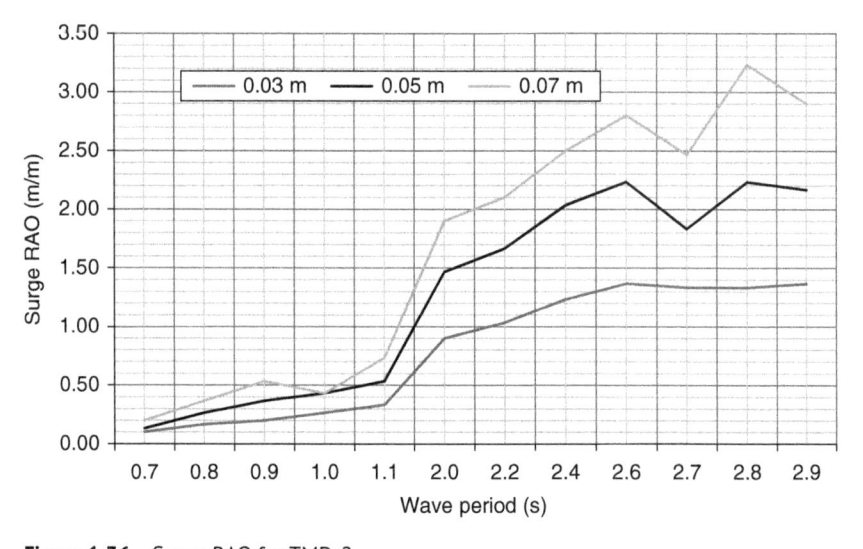

Figure 1.36 Surge RAO for TMD-2.

1.11.2 MLAT with a TMD

The scale model of the MLAT is now investigated in response to regular wave impacts in the presence of a TMD with three chosen configurations, as discussed previously. Figures 1.35–1.37 show surge RAOs for the three TMD configurations, respectively. It is seen that the MLAT fitted with TMD-1 exhibits a response

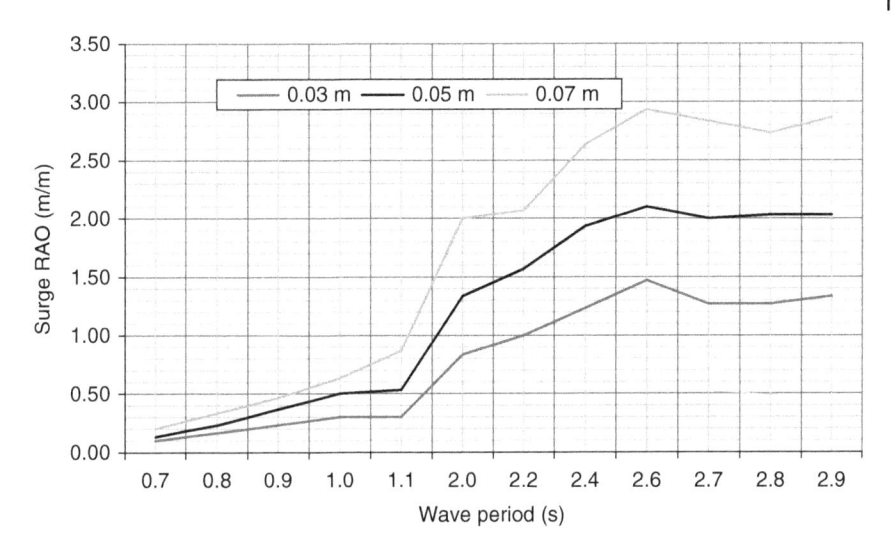

Figure 1.37 Surge RAO for TMD-3.

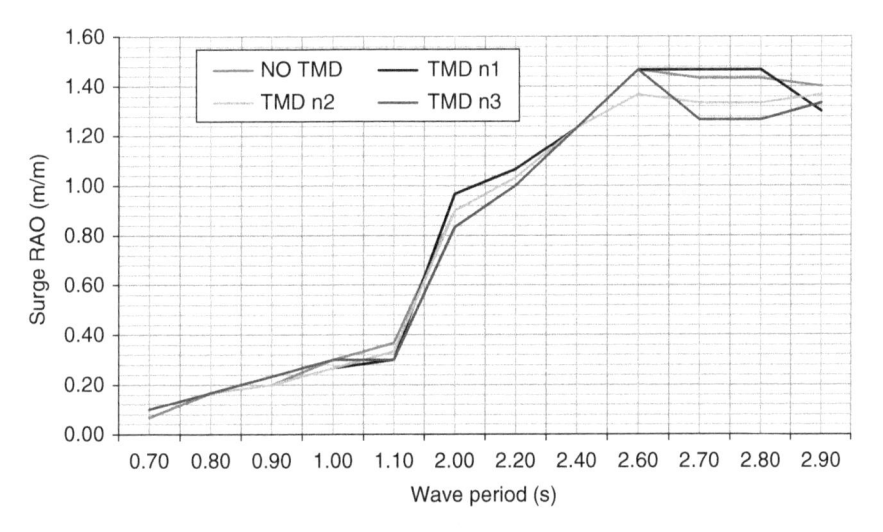

Figure 1.38 Comparison of RAOs for 3 cm wave height.

similar to that in the absence of a TMD; but for other TMD models, response reduction is evident.

In the case of the tower fitted with the TMD-3 model, response reduction is at a maximum, beyond which a TMD will not be effective. Figures 1.38–1.40 compare the response of the MLAT given regular waves 3, 5, and 7 cm high, respectively, for

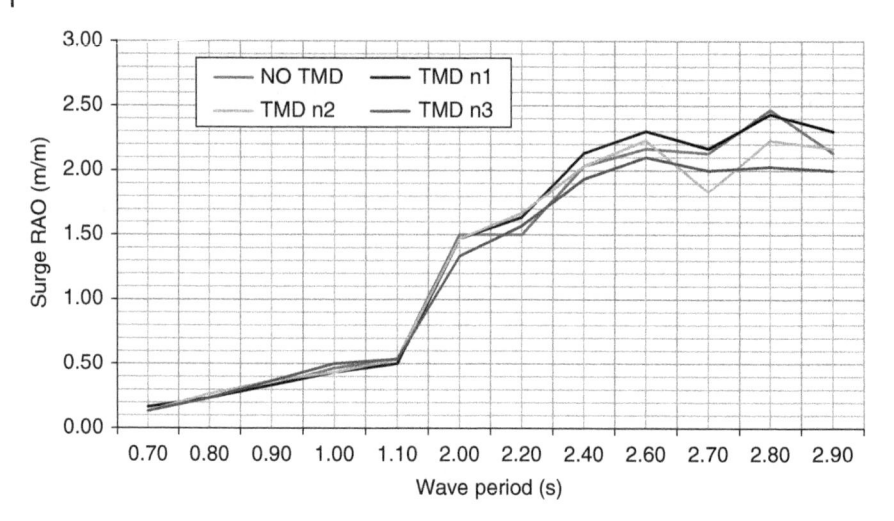

Figure 1.39 Comparison of RAOs for 5 cm wave height.

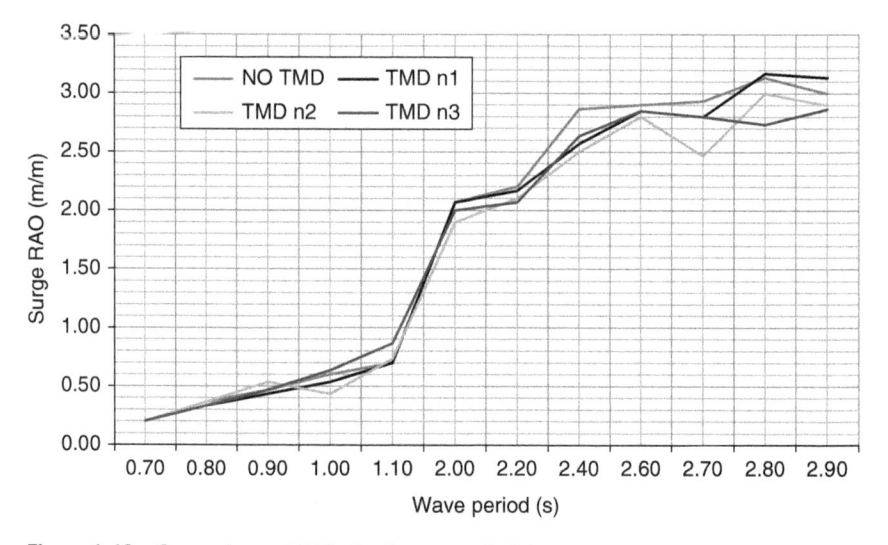

Figure 1.40 Comparison of RAOs for 7 cm wave height.

various periods. It is seen that a TMD attached to the primary system shows an explicit reduction at the period closer to the wave frequency. This shows the effectiveness of tuning the frequency of the damper closer to the resonant frequency band. However, in a case of inappropriate tuning, the response can shoot up, as seen in the case of the tower fitted with TMD-1. However, with the increase in wave height, the response of the tower increases but still depicts the effectiveness of control in the presence of a TMD.

2

Buoyant Leg Storage and Regasification Platforms

Summary

Environmental loads encountered by offshore compliant structures are more severe in deep water in addition to the complexities that arise during their installation. Because existing platforms show serious limitations in terms of storage space, geometric forms of offshore compliant platforms require special attention. A recent development in offshore deepwater platforms is the buoyant leg storage and regasification platform (BLSRP) to store and process liquefied natural gas (LNG) offshore. One of the main operational requirements of LNG tankers is that the degree of compliancy on the topside should be restrained to a large extent. The conceived structural form is a hybrid concept, which restrains the transfer of both rotational and translational responses from the buoyant legs to the deck and vice versa. The proposed platform consists of a deck connected to six buoyant legs through hinged joints; the buoyant legs are connected to the seabed using taut-mooring tethers. Taut-moored tethers and deep-draft buoyant legs resemble the behavior of a tension leg platform (TLP) and spar platform, respectively. The novelty of the design lies in the deployment of the buoyant legs, which are isolated from the large deck by hinged joints. One of the primary advantages is improved functionality in terms of an increase in the storage and processing facilities for LNG. The deck is connected to each of the buoyant legs with separate hinged joints. Because the buoyant legs are not interconnected, independent movement of the legs does not compromise the high degree of compliance offered by the hinged joints. This chapter presents experimental investigations carried out on a scale model of a BLSRP in the presence of regular waves. Numerical studies carried out on the prototype of the BLSRP with both regular and random waves for different wave approach angles are also discussed in detail. The BLSRP shows desirable responses given both operational and phenomenal sea conditions,

---

Offshore Compliant Platforms: Analysis, Design, and Experimental Studies,
First Edition. Srinivasan Chandrasekaran and R. Nagavinothini.
© 2020 John Wiley & Sons Ltd.
This Work is a co-publication between John Wiley & Sons Ltd and ASME Press.

ensuring safe operability and incorporating the main advantage of improved functionality in terms of increasing the storage and regasification capacity for LNG. This new-generation geometric form for the offshore compliant platform is prima facie in favor of the design and development of offshore processing and storage of LNG, which will reduce the cost of oil and gas exploration. Studies carried out on the BLSRP model are contributions from R.S. Lognath (2017, "Dynamic Analyses of Buoyant Leg Storage & Regasification Platforms Under Environmental Loads," PhD thesis submitted to IIT Madras, India) and sincerely acknowledged.

2.1 Background Literature

The installation of offshore platforms in deep water is challenging due to the operations that must be carried out in a hostile environment (Schwartz 2005). Newly evolved geometric designs for such platforms focus on easy installation and simple foundation systems as essential prerequisites (Sohn et al. 2012). A buoyant leg storage and regasification platform (BLSRP) consists of a deck supported by buoyant legs. These legs are position-restrained by tethers with high initial pre-tension. The buoyant legs are subjected to waves and current while the superstructure of the platform is predominantly under the influence of wind. Ball joints placed between the deck and the legs act similarly to hinged connections, restraining the transfer of rotation from the legs to the deck but ensuring a monolithic action by enabling the transfer of displacements. The structural action of the BLSRP under lateral loads is similar to other compliant offshore structures: (i) tension leg platform (TLP), because restraining systems with tethers are common; (ii) spar platforms, because each buoyant leg resembles a spar buoy due to deep-draft conditions; and (iii) articulated towers, due to the presence of hinged joints. Therefore, the BLSRP is a hybrid compliant platform, conceived from the existing successful geometric forms of offshore compliant platforms.

Environmental loads encountered by offshore compliant structures are more severe in deep water in addition to the complexities that arise during their installation. Because existing platforms show serious limitations in terms of storage space, geometric forms of offshore compliant platforms require special attention (ABS 2014). The world's energy demand is rising rapidly, and alternative resources are being explored to match the deficit; liquefied natural gas (LNG) is one of the alternatives being explored (API 2005). Recent studies show that transporting LNG for long distances to shore imposes a heavy penalty in terms of cost and environmental issues (Lloyd's Register 2005). It is therefore imperative to reduce the transport cost by processing LNG offshore by deploying large storage and regasification units adjacent to the production wells. Based on this pursuit, a new offshore compliant structure – the BLSRP – is proposed to store and process LNG offshore. The proposed platform consists of a deck connected to six buoyant legs through hinged joints; the buoyant legs are connected to the seabed using taut-mooring tethers. The conceived structural form is a hybrid concept that restrains the transfer of both rotational and translational responses from the buoyant legs to the deck and vice versa. Taut-moored tethers and deep-draft buoyant legs resemble the behavior of the TLP and spar platform, respectively.

Large floating, storage, and regasification units (FSRUs) are finding increased applications in the offshore oil and gas fields in recent years (DNV 2010a, b). Apart from the economic advantages, the fabrication and commissioning time is considerably less (two to three years) than an onshore plant import terminal, which takes about five to seven years (ABS 2014). The construction of loading and receiving LNG terminals requires huge investments, and launching an FSRU is safer (DNV 2011). This chapter describes the preliminary geometric design of a typical storage and regasification platform consisting of a regasification unit, a gas turbine with a generator, air compressors, fuel pumps, a fire water and foam system, a freshwater system, cranes; a lubricating oil system, lifeboats, a helipad, and a LNG tank on the deck. The novelty of the design lies in the deployment of buoyant legs, which are isolated from the large deck by hinged joints.

The advantages of hinged joints on compliant offshore structures have been well demonstrated by various researchers in the recent past (Chandrasekaran and Madhuri 2015). Hinged joints restrain rotational motion from the buoyant legs to the deck, which enables better recentering under wave loads (Chandrasekaran 2015a, b). On the other hand, the larger response of the deck in compliant degrees of freedom (DOF) such as surge, sway, and yaw under wind loads does not cause additional rotation in the buoyant legs due to the presence of the hinged joints. Detailed dynamic analyses of the proposed platform under environmental loads are scarce in the literature. While large floating LNG carriers are recommended to operate in sea conditions 4–6, as measured on the Douglas sea scale (World Meteorological Organization 2014), it is proposed to investigate the suitability of a BLSRP given sea conditions 5–8 (ClassNK 2015), because it is taut-moored (Shaver et al. 2001; Capanoglu et al. 2002). One of the main operational requirements of LNG tankers is that the degree of compliancy on the topside should be restrained to a large extent (Chandrasekaran 2016a, b).

The current study discusses the design of a new and innovative geometric form whose hull is isolated from the supporting legs. The deck of the BLSRP is supported by buoyant legs, which are similar to those of a tethered spar with a single or group of cylindrical water-piercing hulls. Buoyant legs are alternative structural forms of spar platforms, as they are positively buoyant with a deep draft. The proposed platform consists of a deck connected to six buoyant leg structures (BLSs) through the hinged joints. The buoyant legs are connected to the seabed using taut-mooring tethers. The conceived structural form is a hybrid concept that restrains the transfer of both rotational and translational responses from the buoyant legs to the deck and vice versa. One of the main advantages is improved functionality in terms of an increase in the storage and processing facilities for LNG.

2.1.1 Buoyant Leg Structures

Graham and Webb (1980) addressed the design of a tethered buoyant platform production system, highlighting the technical and economic advantages in rough sea conditions such as the North Sea. Halkyard et al. (1991) implemented a coupled in-place analysis of a tethered buoyant tower (TBT) using the finite element software COPIPE under wind, current, and wave drift loads. Wind loads were estimated using American Bureau of Shipping (ABS) rules, and the wave loads were estimated using diffraction theory. Experimental verification was also performed on a 1 : 89 scale model for wave loads only. Based on the studies, it was concluded that the TBT was a more cost-effective structure than conventional offshore structures. Perryman et al. (1995) explained the concept of a TBT for hydrocarbon reservoir operations, capable of supporting up to 18 wells at a water depth of 1800 ft. Installation, operation, the riser effect, and cost estimates were addressed. The tether tension response amplitude operators (RAOs) of the TBT were observed to be less than that of a TLP. Response analyses of an offshore triceratops supported by a buoyant leg structure showed stable responses given operational sea conditions; ball joints compromise for the large rotational displacements of the deck (Chandrasekaran and Madhuri 2012, 2015). Form-dominated offshore structures exhibit satisfactory performance under the encountered loads while improving the high degree of compliance (Chandrasekaran 2014, 2015a, b, 2016a, b; Stansberg et al. 2004).

Copple and Capanoglu (1995) projected a cost-effective field development concept called a BLS that is simple to fabricate, transport, and install, and also exhibits relatively reliable in-service performance characteristics for offshore operational requirements and environmental criteria. The cost data, including a comparative assessment of alternative concepts, were discussed, and the advantages were highlighted. Shaver et al. (2001) conducted experimental and numerical responses of a tethered buoyant leg structure in the presence of regular waves. The BLS comprises of eight cylindrical water-piercing structures connected spherically and forming a moon pool for drilling purposes. Diffraction analysis has been performed using WAMIT by imposing tether stiffness as well as pre-tension. Parametric studies were conducted to document the sensitivity of motion responses to various parameters, including tether tension, tether stiffness, and buoyant leg rotational stiffness. The experimental and analytical results had a discrepancy in the surge, pitch, and tether tension and the RAOs of the moon pool (Wu et al. 2014). Capanoglu et al. (2002) investigated comparison studies of the model test and motion analysis of the BLS for various loads arising from waves, wind, and currents. Analytical studies were performed using the WAMIT software. Modeling of the BLS was carried out using the Froude scale. The experimental and analytical results had some discrepancies. The BLS hull structure may be modeled using a Morison equation loading on a stick model or a diffraction solution. Both modeling methods yield similar results in time-domain or frequency-domain simulations. Nonlinear effects become important in time-domain simulations and appear to produce better approximations to measured data (Stansberg et al. 2002; Buchner et al. 1999).

2.1.2 Floating Production and Processing Platforms

Brown and Marvakos (1999) conducted a comparative study of the dynamic analysis of suspended wire and chain mooring lines used in floating production systems (FPSs). The analytical results were based on time or frequency domain methods for a chain mooring line suspended in shallow water and a wire line in deeper water. They recommended bi-harmonic top-end oscillations representing combined wave and drift-induced excitation, which is also an alternate method of analysis (Nielsen and Bindingbø, 2000). Hwang et al. (2010) examined the scheme of floating production, storage, and offloading (FPSO) field developments, design procedures, and activities of the hull of an FPSO unit installed in the near offshore area of Nigeria. They presented tension variations in mooring lines and damping due to viscous drag forces under the assumption that the magnitude of mooring tension is relatively more in comparison to its submerged weight. The method was found to be suitable for estimating damping at early design stages for deepwater taut-mooring systems of FPSOs and FPSs, etc. This method also estimates the natural periods of the mooring and towing lines while including dynamic effects caused by extreme loads or damping. The proposed method predicts extreme loads and damping to better accuracy when the amplitude of oscillation is moderate.

Zhao et al. (2013) experimentally examined the effects of inner-tank sloshing on the responses of floating liquefied natural gas (FLNG) and found that it is sensitive to wave excitation frequencies. Given a roll motion, specifically, inner-tank sloshing can affect the wave frequency motions of the vessel, whereas it is insignificant for low-frequency motions. A comparison of the three filling conditions was used to examine the influence of filling levels. Paik et al. (2011) investigated possible changes in the structural design of the membrane corrugations in cargo tanks, which essentially requires an exact analysis of strong performance in a membrane-type LNG carrier cargo containment system under cargo static pressure loads in a cryogenic condition with a temperature of $-163\,°C$ and large nonlinear elastic–plastic deflection. The finite element analysis (FEM) used ANSYS to analyze structural behavior under cargo pressure loads. Zhao et al. (2011) presented a comprehensive review of recent research developments on the hydrodynamics of FLNG. Research results based on numerical calculations and model tests are summarized, existing problems are discussed, and further research topics regarding FLNG are suggested. Further investigations should be undertaken on the excitation mechanisms of the motions of the fluid enclosed between the platform and carrier and the hydrodynamic interactions between the bodies and the enclosed fluid. Future work should concentrate on improving hydrodynamic modeling, including the attainment of more accurate time-domain radiated wave forces and the consideration of nonlinear and viscous effects.

2.2 Experimental Setup

This section describes the experimental setup used to carry out investigations on scale models of a BLSRP in the presence of regular waves. A scale ratio of 1 : 150 is chosen to suit the wave flume water depth of 4.0 m with a prototype water depth of 600 m. Froude scaling is chosen to ensure the scaling of gravity forces and wave resistance in the model. Geometric, kinematic, and dynamic similarities are considered for scaling the geometric size, hydrostatic stiffness, tether stiffness, mass properties, and hydrodynamic loading. Scale models are examined in the presence of regular waves with specific wave heights and different wave periods. Experimental investigations are carried out in a deepwater wave flume 90 m × 4 m × 4.5 m deep. The beach end of the flume consists of rubble-mound filling over a length of about 7 m. The wave generator has two flaps that can be moved independently by a servo-hydraulic system. Figure 2.1 shows the setup.

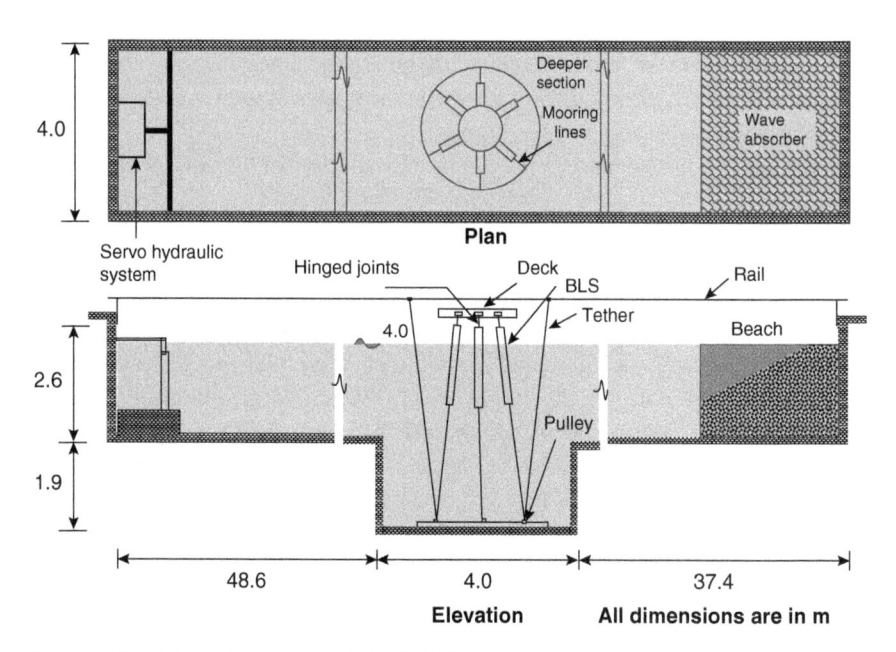

Figure 2.1 Schematic diagram of the BLSRP installed in a wave flume.

Piezoelectric accelerometers are used to measure translational responses, while inclinometers are used to measure rotations. Dynamic tether tension variations are measured using ring-type load cells, which are fabricated from stainless steel. The external diameter of the load cell is 32 mm, the internal diameter is 28 mm, and the thickness is 2 mm. Foil strain gauges protected by epoxy resin are used with a half-bridge configuration. The wave surface elevation of the incident wave is recorded using a resistant-type wave probe. Figure 2.2 shows the arrangements for the experimental setup – the deck is connected to each buoyant leg with separate hinges. The structural details of the BLSRP are summarized in Table 2.1.

2.3 Experimental Investigations

Accelerometers are fixed on the top of each buoyant leg and the deck to measure translations as well as an inclinometer to measure rotations, respectively. Mooring lines are connected to each leg at 20° with respect to the base plate with taut-mooring lines whose stiffness is adjusted by a ratchet mechanism as shown in the figure. The free-floating translation and natural rotational periods of a scale model of the BLSRP are measured experimentally by conducting free decay tests in the respective DOF. The damping ratio in the respective DOF is obtained by the logarithmic decrement method and is given here:

$$\delta = \frac{1}{n} \ln\left(\frac{x_0}{x_n}\right) \quad (2.1)$$

where x_0 is the higher value of the two peaks. x_n is the value of the peak after n cycles. The damping ratio is determined by:

$$\zeta = \frac{1}{\sqrt{1 + \left(\frac{2\pi}{\delta}\right)^2}} \quad (2.2)$$

The damping ratio thus obtained is used to find the natural frequency ω_n of the vibrating system. The natural damping frequency ω_d of the system is given by:

$$\omega_d = \frac{2\pi}{T} \quad (2.3)$$

where T is the period between two successive amplitude peaks. The natural frequency and natural period are given by:

$$\omega_n = \frac{\omega_d}{\sqrt{1-\zeta^2}} \quad (2.4)$$

66 *Offshore Compliant Platforms*

Figure 2.2 Experimental setup and arrangements: (a) side view; (b) hinged joint; (c) roller at the base plate for guiding the mooring line; (d) load cell; (e) ratchet mechanism for adjusting the tension; (f) plan arrangement.

Table 2.1 Structural details of the BLSRP.

Description	Prototype	Units	Model (1:150)	Units
Water depth	600	m	4000	mm
Mass of the structure	400 000	ton	118.51	Kg
Utilities	10 000	ton	2.93	Kg
Secondary deck plate	1250	ton	0.37	Kg
Stainless steel tank	1800	ton	0.53	Kg
LNG	25 000	ton	7.40	Kg
Main deck plate	2500	ton	0.74	Kg
BLS (6)	25 500	ton	7.55	Kg
Ballast	333 950	ton	98.94	Kg
Diameter of the BLS	22.50	m	150	mm
Length of the BLS	200	m	1333.33	mm
Diameter of the deck	100	m	666.66	mm
Draft	163.57	m	1117.73	mm
Meta centric height	15.18	m	114.86	mm
Length of the tether	470.84	m	3138.93	mm
Initial tether tension	8715.89	ton	2.50	Kg
Deck				
I_{XX}, I_{YY}	2 530 725.50	ton-m^2	33 326.42	Kg-mm^2
I_{ZZ}	4 729 752	ton-m^2	62 284.8	Kg-mm^2
r_{XX}, r_{YY}	7.90	m	50.66	mm
r_{ZZ}	10.80	m	72	mm
Single buoyant leg				
I_{XX}, I_{YY}	85 147 115	ton-m^2	1 121 278.81	Kg-mm^2
I_{ZZ}	1 159 825.3	ton-m^2	15 273.41	Kg-mm^2
r_{XX}, r_{YY}	37.70	m	251.33	mm
r_{ZZ}	4.40	m	29.33	mm
Tether diameter	0.05	m	0.33	mm
Tether stiffness	875 741	N/m	0.03	N/mm
Height of the LNG tank	7	m	46.66	mm
Modulus of tether	2.1E+11	N/m^2	1400	N/mm^2

Table 2.2 Natural periods and damping ratios.

	Tethered BLSRP (experimental)			Numerical studies (prototype)			
	Natural period (s)			Free-floating		Tethered	
Description	Model	Prototype	Damping ratio (%)	Natural period (s)	Damping (%)	Natural period (s)	Damping (%)
Surge	9.21	112.76	8.12	—	—	118.50	8.55
Sway	9.26	113.37	8.30	—	—	121.00	8.45
Heave	0.28	3.42	3.50	3.21	3.64	3.18	3.65
Roll	0.33	4.04	6.44	4.15	6.50	—	—
Pitch	0.35	4.34	6.63	4.25	6.60	—	—
Yaw	4.90	59.97	7.12	—	—	—	—

$$T_n = \frac{2\pi}{\omega_n} \quad (2.5)$$

Table 2.2 shows the results of free vibration tests carried out on the scale model and prototype during experimental and numerical studies, respectively. It is seen from the table that natural periods of the BLSRP resemble any typical tethered compliant structure like a TLP but show relatively greater stiffness in the yaw DOF (Srinivasan Chandrasekaran 2015b). Greater stiffness in yaw motion is attributed to the symmetric layout of the buoyant legs, which are independently spread at the bottom but closely connected to the deck. As hinged joints are under high axial force imparted by tethers, their moment-rotation characteristics vary compared to their behavior in the absence of axial force. It is very complex to capture this time-dependent behavior and incorporate it in the numerical model, which is influenced by the variable buoyancy of the legs. Discrepancies between the experimental and numerical studies, as seen in the estimate of natural periods and damping ratios, are attributed to this effect.

Figure 2.3 shows the orientation of the platform with a 0° approach angle. Figure 2.4 shows the response of the platform in different DOF with a 0° wave approach angle and dynamic tether tension variations. Variation is measured only from the initial pre-tension values. Results are shown for a wave height of 0.1 m, which corresponds to a 15 m wave height in the prototype for the chosen scale of the model. It is seen that the deck response, both in translational and rotational DOF, is less than that of the buoyant legs, indicating greater operational convenience and safety. Similar behavior is also verified for other wave approach angles (Chandrasekaran and Lognath 2015).

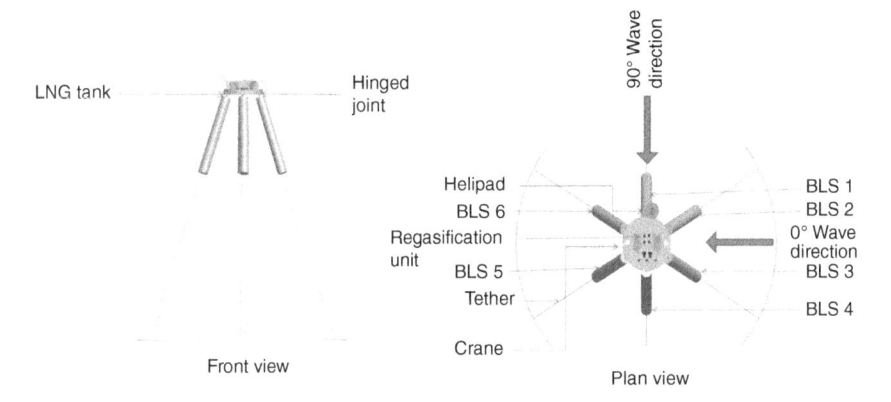

Figure 2.3 Orientation of the BLSRP for the wave heading angle.

It is seen from the figures that the deck response is significantly less than that of the maximum response in all active DOF. It can also be observed that the responses of the deck and buoyant legs are symmetric about the abscissa, with less residue indicating high recentering capabilities. This behavior is attributed to the restraint offered by the hinged joint in both the translational and rotational DOF. Differences in the responses of the buoyant legs are due to the variable submergence effect, which is one of the primary sources of nonlinearity in the excitation force. The presence of hinged joints at each BLS unit isolates the deck from the legs and thus improves operator comfort and the safety of the platform. The presence of rotational responses in the deck, despite the presence of hinged joints, is due to the differential heave response that occurs due to the dynamic tether tension variations. As buoyant legs are symmetrically spread with respect to the vertical axis of the platform, it is imperative to envisage a non-uniform phase lag in the recentering process; this causes the yaw motion on the deck. The roll response of the deck is about 60% less than the maximum response of the buoyant leg, while the heave response is about 40% less. The rotational responses of the deck are less than those of the buoyant legs, but this still remains a design challenge that needs detailed investigation. Table 2.3 shows the maximum values of the response of the BLSRP given a (0°, 0.1 m) wave heading combination. By comparing the response amplitudes, it is seen that the deck response is appreciably lower than the maximum response of the buoyant legs in all DOF.

Greater stiffness in yaw motion is due to the symmetric layout of the buoyant legs, which are spread at the bottom. The deck response, which is significantly less than that of the buoyant legs, validates the use of the hinged joints; they do not transfer rotations from the legs to the deck. A lower heave response of the deck, compared to that of the BLS units, ensures comfortable and safe operability.

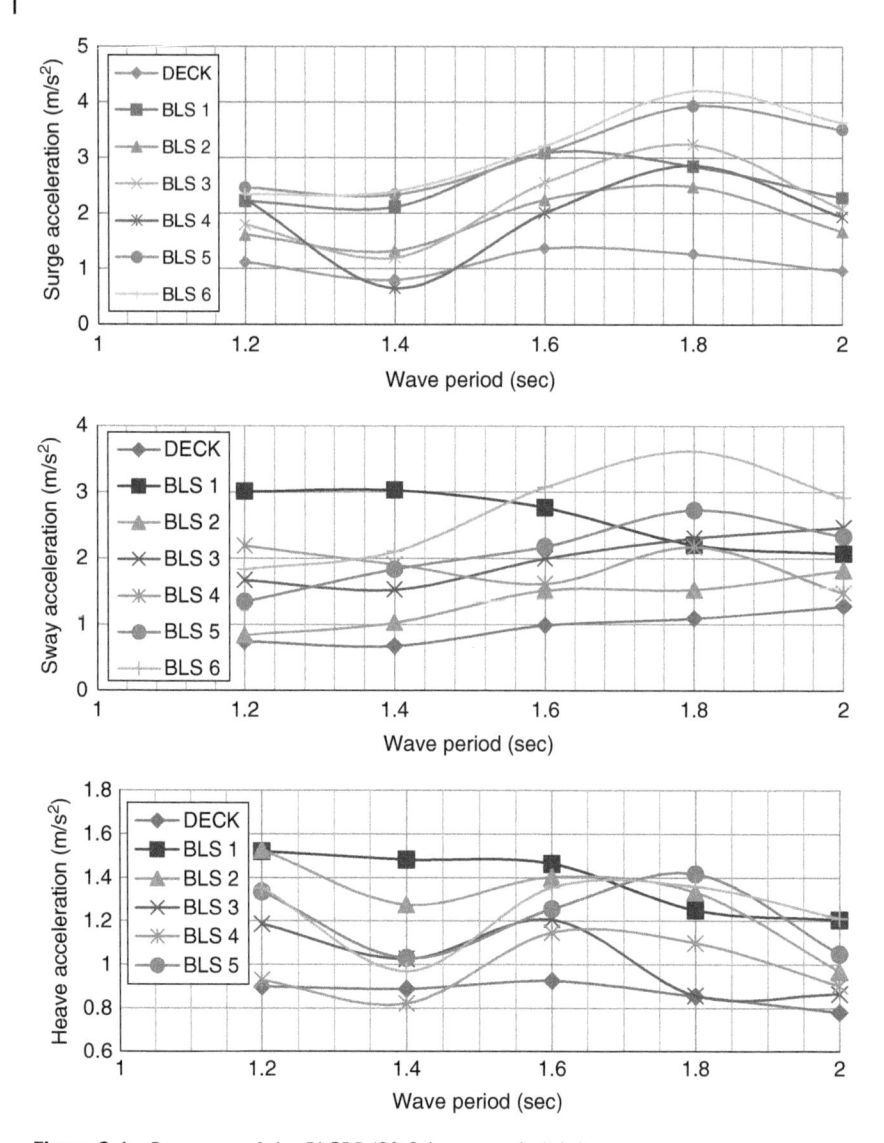

Figure 2.4 Response of the BLSRP (0°, 0.1 m wave height).

Hinged joints also serve as isolators, which control the deck motion even for a large movement/rotation of the buoyant legs. It is also interesting to note the significant difference in the response of different buoyant legs in various DOF. This is mainly due to their asymmetrical positions with respect to the incident wave direction. Further, the deck response in the rotational DOF is exceptional due to the differential heave in various buoyant legs. This also induces dynamic tether

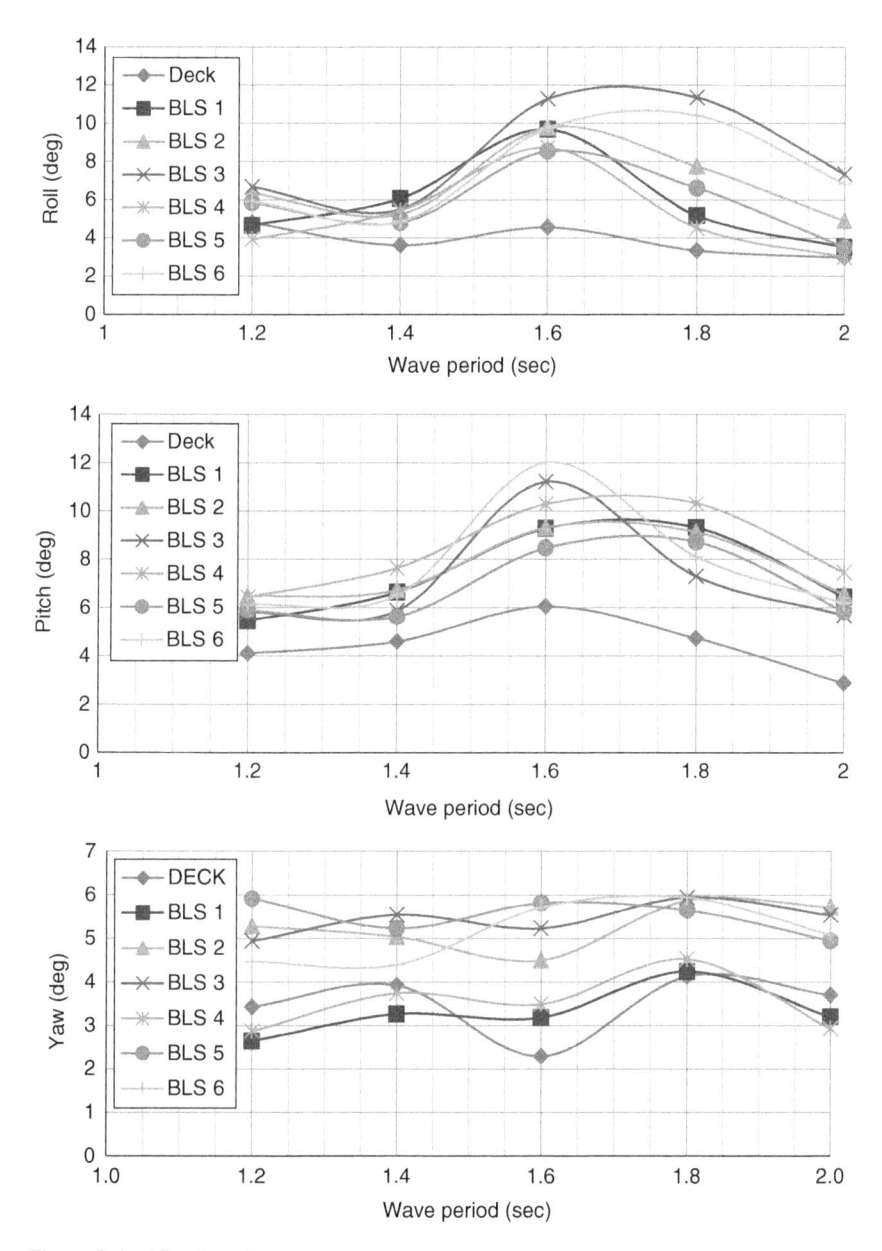

Buoyant Leg Storage and Regasification Platforms | 71

Figure 2.4 (Continued)

Table 2.3 Maximum response of the BLSRP model (0°, 0.1 m).

Description	Deck	Leg 1	Leg 2	Leg 3	Leg 4	Leg 5	Leg 6
Surge (m)	1.36	3.09	2.48	3.23	2.86	3.93	4.20
Sway (m)	1.27	3.03	1.82	2.46	2.18	2.73	3.62
Heave (m)	0.92	1.52	1.53	1.21	1.15	1.42	1.36
Roll (deg)	4.77	9.68	9.78	11.34	8.72	8.53	10.41
Pitch (deg)	6.05	9.31	9.31	11.22	10.31	8.72	12.02
Yaw (deg)	4.13	4.24	5.87	5.94	4.52	5.91	5.94

tension variations in the tethers, which need to be assessed for Mathieu instability (Chandrasekaran and Kiran 2018).

The geometric design of the buoyant legs, as attempted in this novel platform design, is a recent advancement, which successfully exhibits a good recentering capability for the deck in all of the translational DOF. The yaw response of the deck is attributed to the time delay in the recentering capability of the buoyant legs under directional wave loads. As the buoyant legs are spread with respect to the axis of the deck, it is imperative to envisage a non-uniform phase lag in the recentering process, which results in the yaw motion of the deck. It is also seen that the maximum tether tension variation is about 20%, which is within permissible limits. The presence of hinged joints at each buoyant leg partially isolates the deck from the legs, improving both operator comfort and the safety of this novel geometric form of offshore compliant platform.

2.4 Numerical Studies

A detailed set of numerical investigations are also carried out on the BLSRP. The numerical model of the new BLSRP was developed using Creo Elements/Pro software. Major components of the model are the deck, LNG tank, buoyant legs, hinged joints, and mooring lines. The deck is connected to the legs through hinged joints, and the buoyant legs are moored to the seabed by taut-moored tethers. Six strands are used as mooring lines in the numerical model. The hydrodynamic problem is solved using three-dimensional panel methods. The software used in the study can simulate fluid–structure interactions in all six DOF based on potential flow theory, while the panel method is used to derive the hydrodynamic loads. Radiation/Diffraction theory is employed, which uses the characteristic dimensions that cause scattering of the incident regular waves to obtain both the first- and second-order wave loadings on the floating bodies. While evaluating Morison

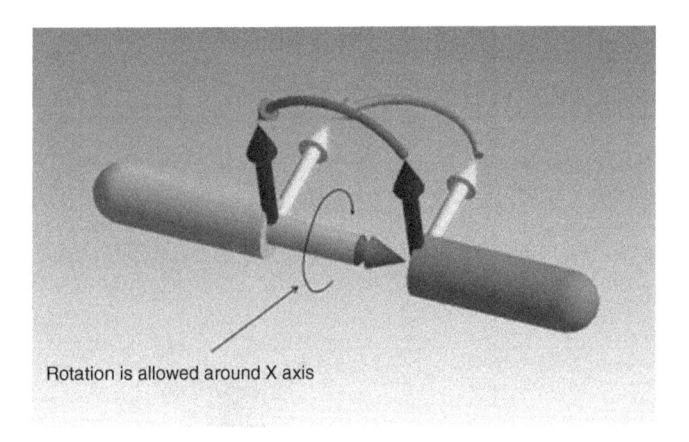

Figure 2.5 Details of the hinged joint in the numeric model.

loads on such components of the structure, several factors are considered. Incident flow is expected to be modified by the presence of the main structure due to diffracted wave forces. Water particle kinematics are updated accordingly to account for this effect. The local water surface during the passage of waves is also modified due to the presence of the structure; this affects wave loading on the structure. The overall increase in wave height, called the *caisson effect*, and the dynamic recentering capabilities of the platform, called the *ride-up effect*, are also accounted for in the numerical model. Linear wave theory is considered sufficient in evaluating the incident and diffracted wave effects. Tethers are modeled as linear, elastic cables. The dynamic tether tension variation is checked at every instantaneous position of the platform. Stiffness variations in the platform geometry arising from the instantaneous position of the platform are automatically updated in the dynamic analysis. The dynamic equilibrium position of the platform is assessed given various nonlinearities that arise from the encountered lateral loads, varying tether tension, and variable submergence effects. Figure 2.5 shows the details of the hinged joint.

The BLSRP is analyzed numerically for different wave approach angles. Figure 2.6 shows the response of the platform with a 30° wave approach angle and 15 m wave height. Figure 2.7 shows the dynamic tether tension variations in the mooring lines.

It is seen from the figures that both the surge and sway surge responses of the deck are minimal in comparison to those of the buoyant legs. It can also be observed that the deck response in the rotational DOF is less because hinged joints restrain the transfer of rotation from the legs to the deck. However, the roll and pitch responses of the deck are due to the cumulative effects of differential heave of the deck given wave action. This is seen from the tether tension variation of each

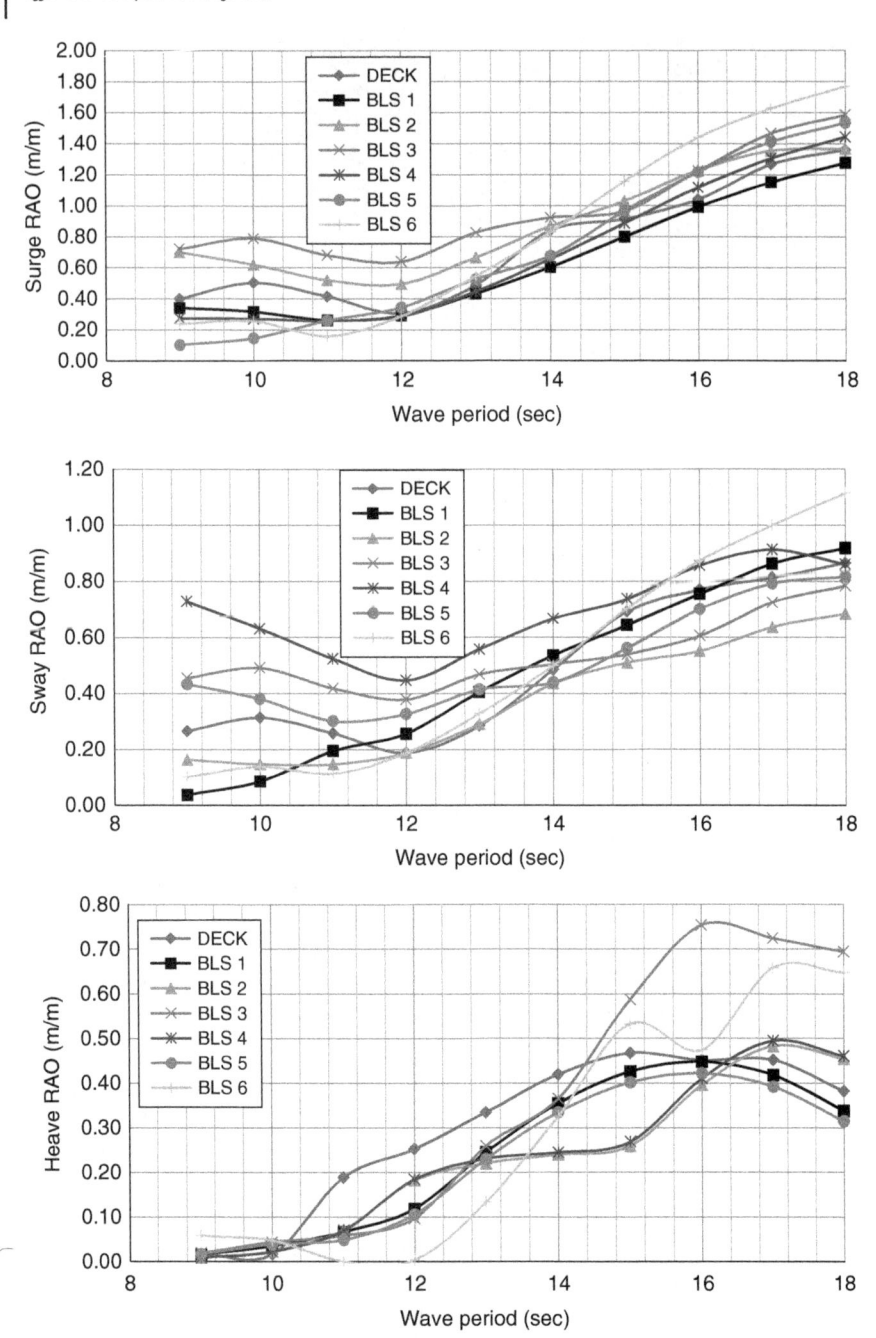

Figure 2.6 Response of the BLSRP (30°, 15 m).

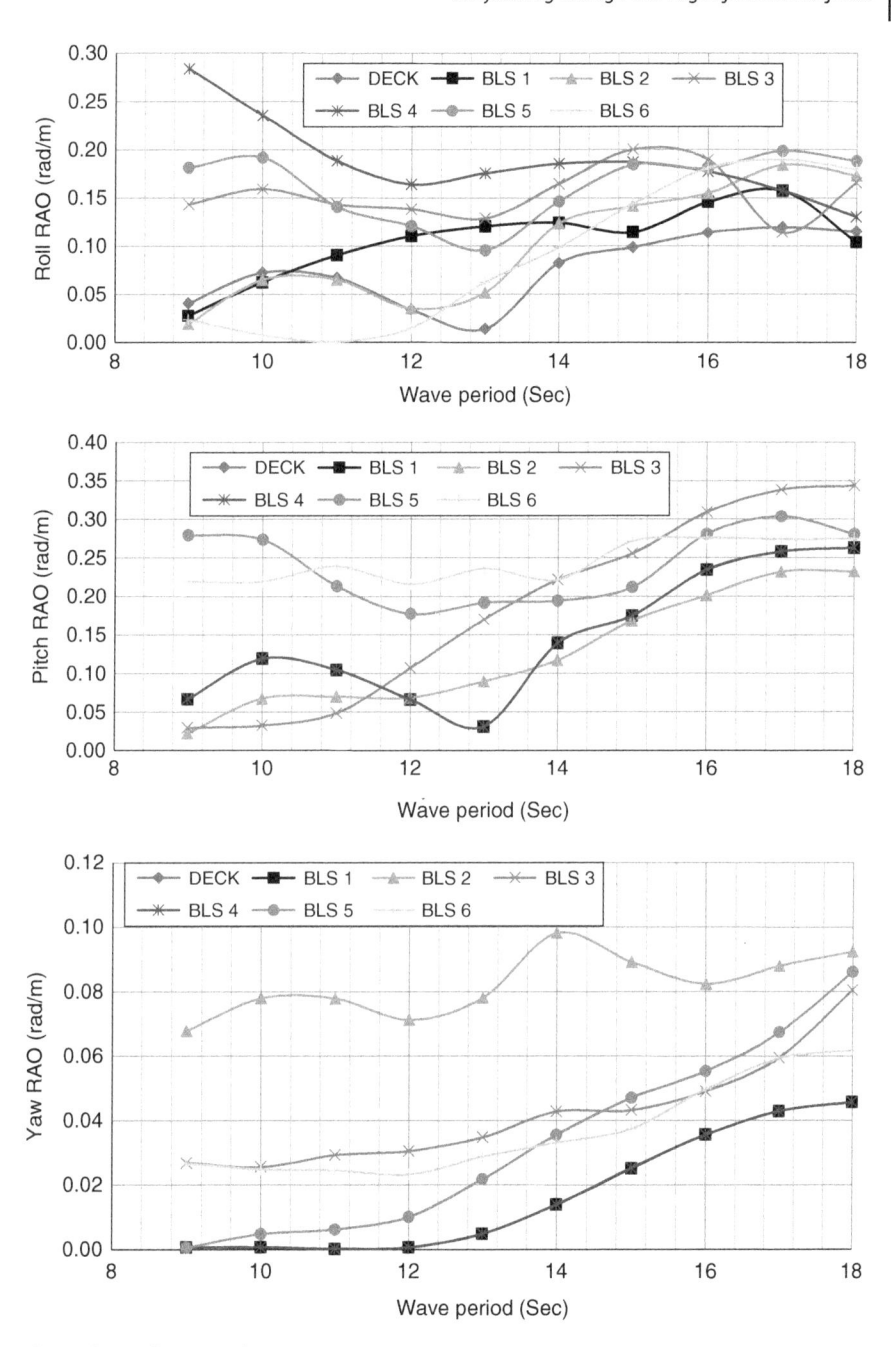

Figure 2.6 (Continued)

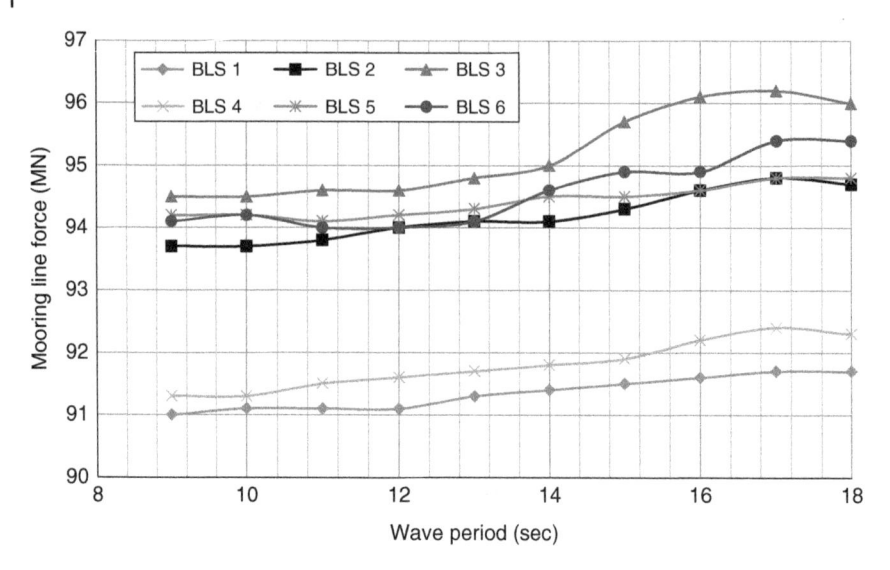

Figure 2.7 Tether tension variations in mooring lines.

buoyant leg. Variations in the tension variation also cause roll and pitch motions of the deck. The yaw response of the deck is less than that of the buoyant legs, verifying that the platform exhibits stiff behavior in yaw motion, unlike other taut-moored platforms like TLPs. For the chosen mooring configuration, the platform remains symmetric in the presence of induced waves. Wave loads applied on the buoyant legs induce a response in the deck, validating the fact that the hinged joint imposes a monolithic connection between the deck and the buoyant legs as well.

2.5 Critical Observations

Ultra-deepwater oil and gas exploration increases the necessity for a stable offshore platform that is capable of withstanding the encountered environmental loads while maintaining its operability. Significant displacement of the deck not only damages the platform's structural system but also creates an uncomfortable environment for people working on-board. Also, it raises complexities regarding the subsea pipeline connected to the platform. A new-generation platform, the BLSRP, showed satisfactory behavior in response to both directional waves and critical sea conditions. Experimental investigations carried out on a scale model of the platform showed that for a 0° wave approach angle, the deck response is less than that of the buoyant legs in the translational DOF. A response reduction is also seen in the rotational DOF. In particular, experimental studies show that the surge response of the deck is less than that of the buoyant legs; the maximum reduction is seen in comparison with the response of BLS 6, while this reduction is about 65% in the sway DOF. The heave response of the platform is also controlled due to the presence of hinged joints. The heave response of the deck is about 40% less than that of the buoyant legs; the maximum reduction is seen for the response of BLS 2. The roll response of the deck is less than that of the buoyant legs by about 58%; the maximum reduction is for BLS 4. The reduction in the pitch and yaw responses of the deck is about 49 and 30% for BLS 3 and BLS 6 respectively. For a 90° wave approach angle, the surge and sway responses of the deck are 52.38 and 71.63% less than the response of BLS 6, respectively. It is important to note that the response reduction is seen in the heave response of the deck for various wave approach angles, which is attributed to the presence of the hinged joints. Hinged joints, while ensuring monolithic action between the legs and the deck, are responsible for preventing the transfer of the response from the buoyant legs to the deck in the rotational DOF. The response reduction in heave is due to the reduction in the pitch and roll responses in addition to the high stiffness attributed to the platform, which is achieved by the geometric form-dominated design. The heave response of the deck is about 55.07% less than that of the buoyant legs; the maximum reduction is seen for BLS 4. The roll response of the deck is less than that of the buoyant legs by about 70.61%; the maximum reduction is for BLS 3. The reduction in the pitch and yaw responses of the deck is about 47.74 and 35.97% compared to BLS 3, respectively. The overall tether tension variation in the experimental analysis in the 0 and 90° wave directions is about 20%.

Based on the results of the numerical studies, it is observed that the deck response is less than that of the buoyant legs in both the translational and rotational DOF. This confirms the observations made in the experimental investigations as well. In particular, the heave response of the deck is 29.95% less than the response of BLS 2, while this comparison was about 40% during the experimental investigations. The roll response of the deck is less than that of the buoyant legs by about 70%; the maximum reduction is for BLS 4. With a 45° wave approach angle, the surge response of the deck is 24.85% less than the BLS 6 response, while the sway response is reduced by about 13%. The heave response of the deck is about 27% less than that of BLS 3, which is due to the presence of the hinged joints. The roll response of the deck is less than that of the buoyant legs by about 48%; the maximum reduction is for BLS 4. The reductions in the pitch and yaw responses of the deck are about 34 and 52% in comparison to those of BLS 3 and BLS 5, respectively. Overall, the tether tension variation in the numerical analysis for different wave approach angles is about 12%.

Figures 2.8 and 2.9 show the power spectral density plots of buoyant leg-1 and the deck in different DOF for a 0° wave incidence angle, for (Hs, Tz) as (6 m, 10 s). For a 0° wave direction, the power spectral density (PSD) peaks of the deck's

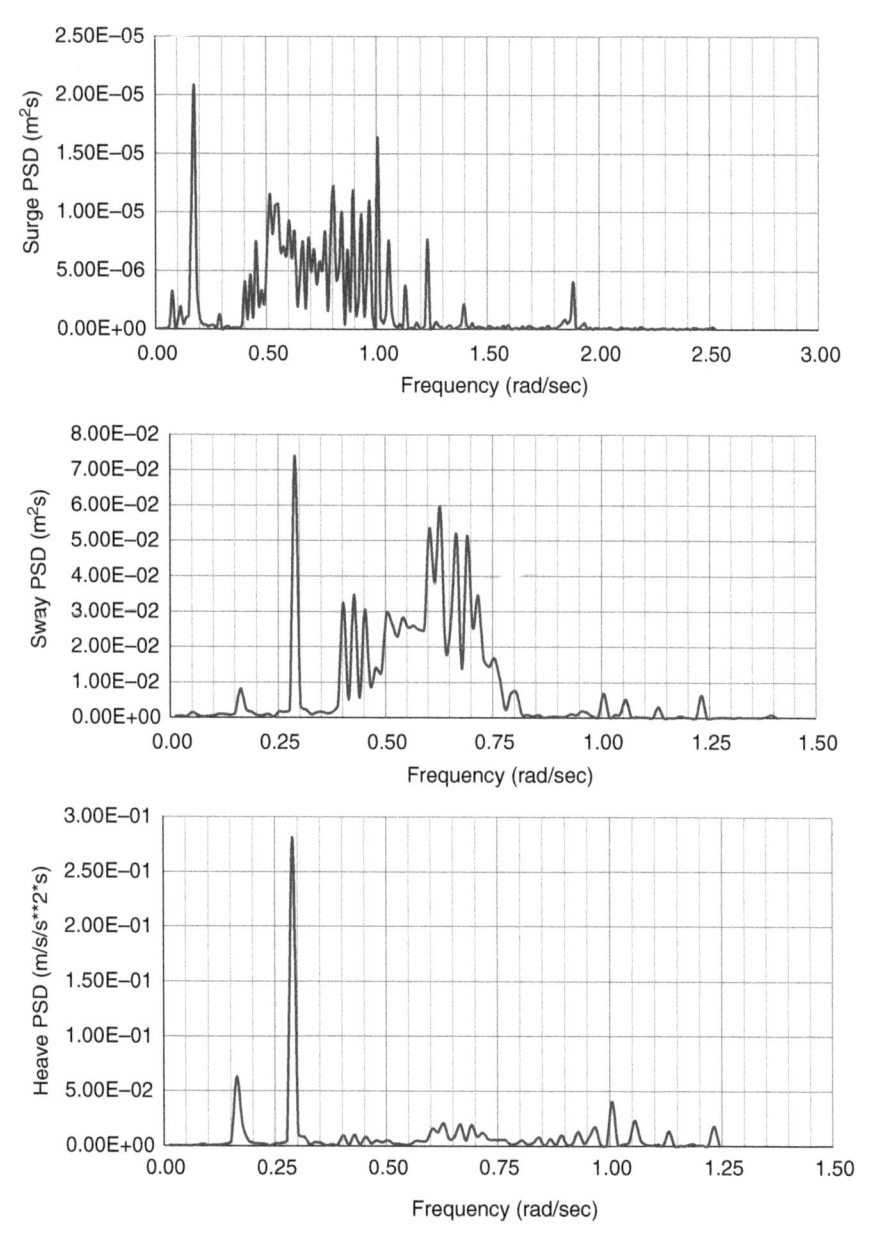

Figure 2.8 Power spectral density plots of buoyant leg 1 (0°, 6 m, 10 seconds).

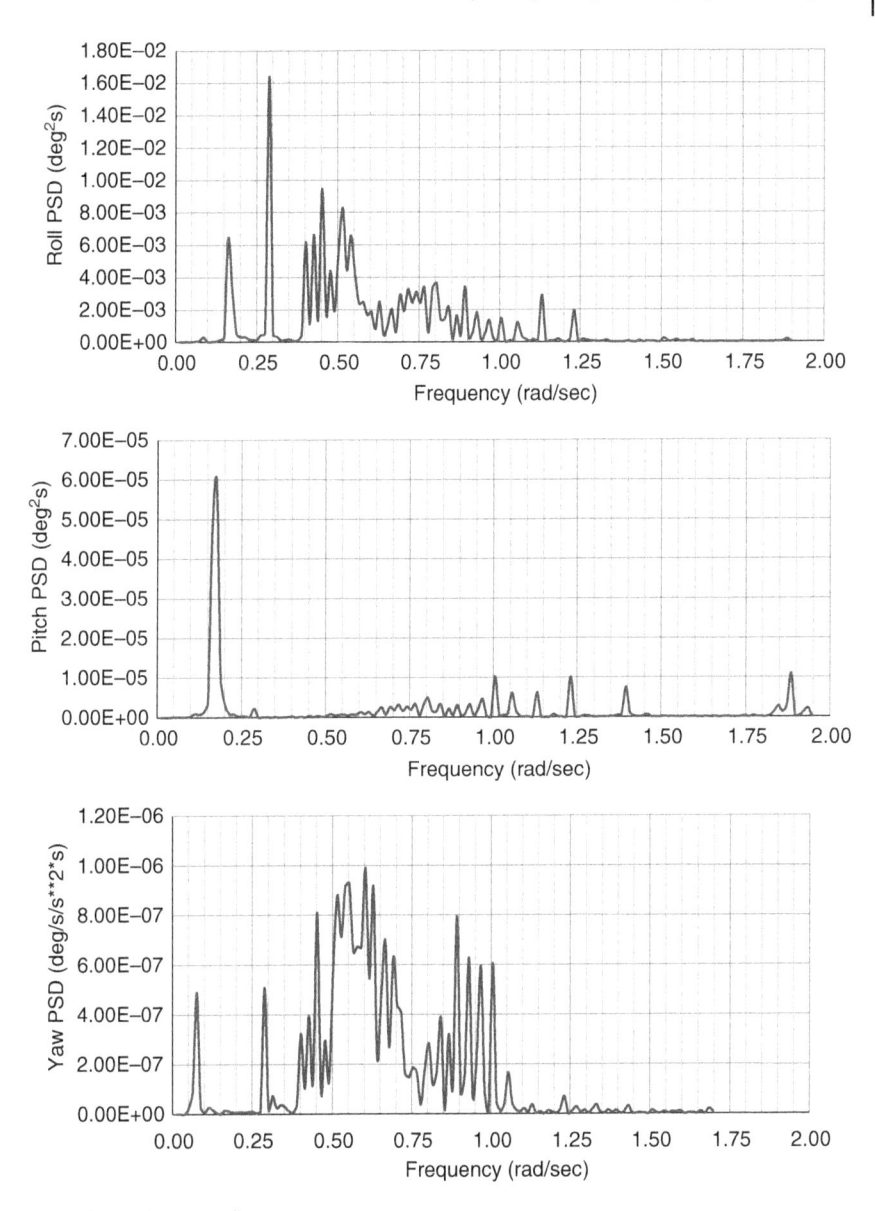

Figure 2.8 (Continued)

80 | *Offshore Compliant Platforms*

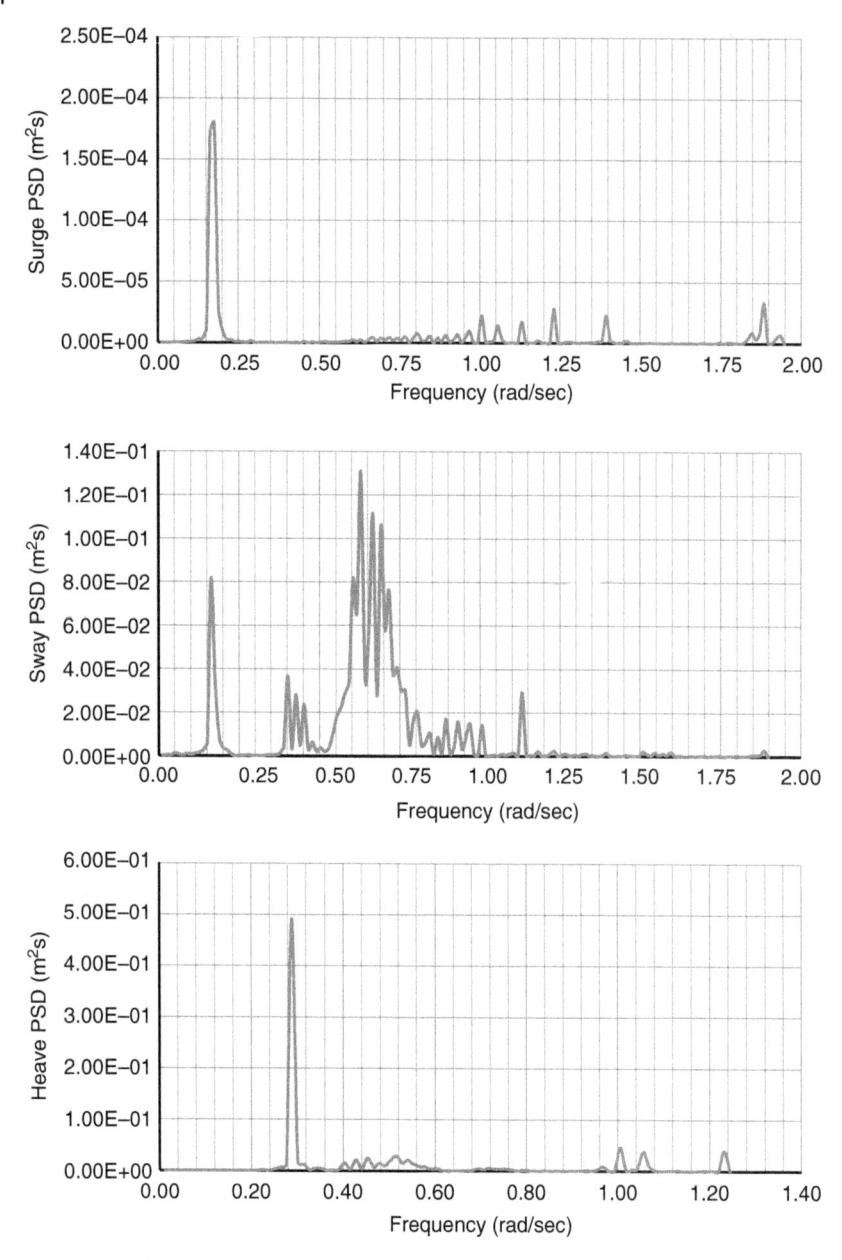

Figure 2.9 Power spectral density plots of the deck (0°, 6 m, 10 seconds).

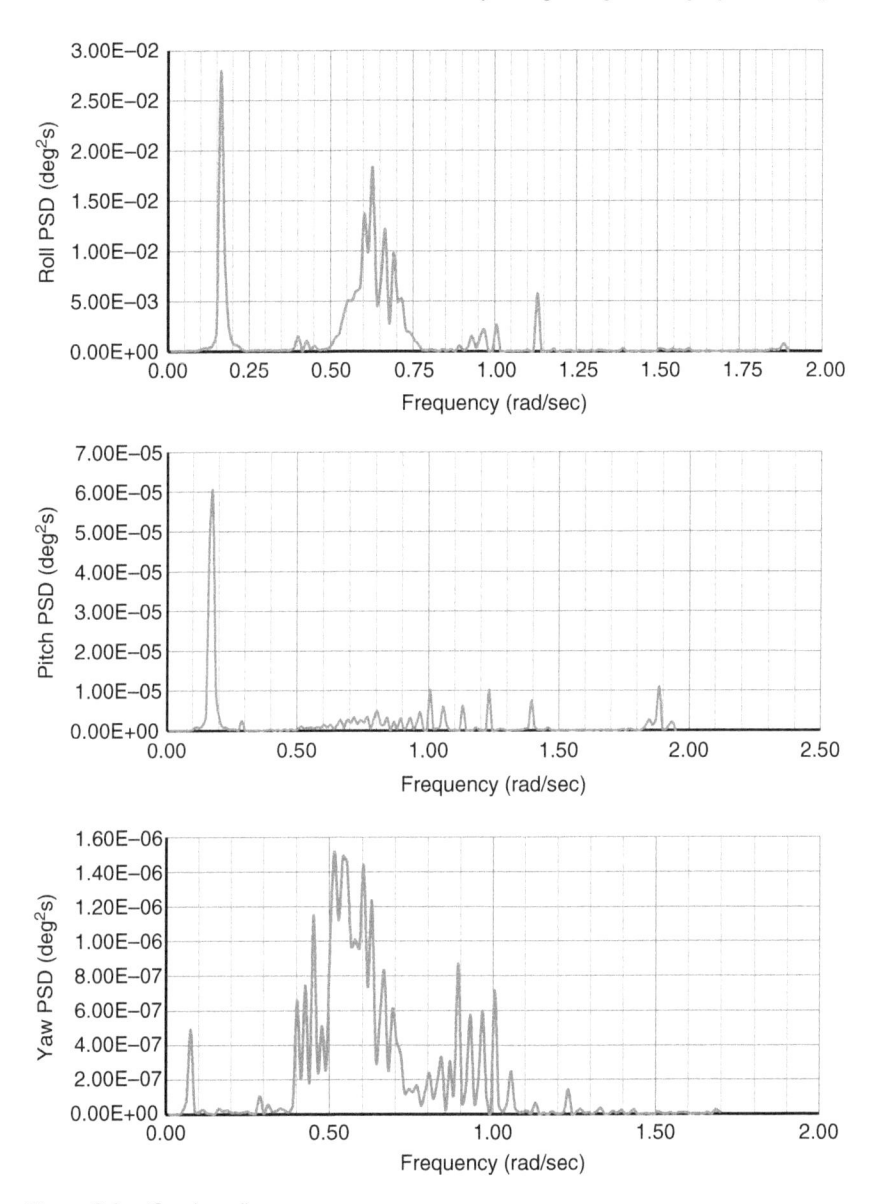

Figure 2.9 (Continued)

surge, heave, and pitch responses are less than those of the highest peak of the buoyant legs. The surge response of the deck showed a wide energy concentration in the range of 0.3–0.6 rad/s. The first peak of the heave response occurs at 0.3 rad/s, which is closer to half of the wave frequency. A small peak also appears closer to the natural heave frequency, while other peaks are widely distributed through the frequencies of the platform in all active DOF. The pitch response plots for the deck showed a significant concentration of energy in the range of 0.18–1.0 rad/s. With a 45° wave direction, the deck response is active in all DOF. The surge and sway responses of the deck showed a wide energy concentration in the range of 0.4–1.0 rad/s. The first peak occurs at 0.4 rad/s, while the second significant peak occurs at 0.8 rad/s. The first peak of the heave response occurs at 0.3 rad/s, which is closer to half of the wave frequency. A small peak also appears closer to the natural heave frequency, while other peaks are widely distributed through the frequencies of the platform in all active DOF. The roll and pitch responses are seen with a wide frequency range of 0.2–1.0 rad/s.

The yaw motion, which is a manifestation of differential heave arising from the tether tension variation, is seen in a narrow frequency range of 0.6–1 rad/s. Phase plots of both the deck and buoyant legs by and large showed that the system is stable and periodic. With a 90° wave direction, the PSD plots of the deck in the surge, pitch, and yaw DOF are insignificant. It is attributed to the presence of a hinged joint, which does not transfer rotational displacement from the buoyant legs to the deck. The PSD of the sway response of the deck shows two significant peaks with a lot of secondary peaks, indicating significant energy in the bandwidth of 0.45–1.1 rad/s. It is also noted that significant peaks occur closer to the excitation frequency of the wave, whereas the first peak occurs at about half of this value. The roll response of the deck shows a similar trend. Although the hinged joint offers a rigid connection between the buoyant legs and the deck, the heave PSD is significantly less than that of the buoyant leg.

Table 2.4 shows the maximum response amplitudes for different wave approach angles for the 6 m wave height, which is the desirable operational sea condition for LNG processing platforms (World Meteorological Organization 2014). By comparing the values, it is seen that the responses in the rotational DOF of the deck are comparatively less than those of the buoyant legs. The presence of hinged joints prevents the transfer of roll and pitch motions from the BLS to that of the deck; however, pitch and roll responses of the deck still arise due to differential heave, as heave is transferred from the BLS to the deck. The yaw response of the deck is less than that of the buoyant legs, which verifies the fact that the platform exhibits stiff behavior in the yaw motion unlike other taut-moored platforms like TLPs. For the chosen mooring configuration, the platform remains symmetric in the presence of induced waves. The roll response of the deck is about 70% less than that of the maximum response of the buoyant legs; the surge and heave

Table 2.4 Maximum response amplitudes (numerical studies; 6 m wave height).

Description			Deck	BLS 1	BLS 2	BLS 3	BLS 4	BLS 5	BLS 6
Surge RAO (m/m)	0°	max	1.64	1.207	1.531	1.554	1.146	1.804	1.793
		min	0.197	0.01	0.319	0.327	0.082	0.119	0.122
	30°	max	1.315	1.253	1.325	1.487	1.321	1.454	1.682
		min	0.179	0.139	0.252	0.405	0.133	0.161	0.211
	45°	max	1.190	0.734	1.181	1.249	0.73	1.073	1.444
		min	0.124	0.02	0.286	0.415	0.238	0.191	0.202
	60°	max	0.912	0.744	0.877	0.916	0.783	0.79	1.105
		min	0.141	0.155	0.211	0.361	0.164	0.171	0.182
	90°	max	0.112	0.042	0.332	0.396	0.03	0.333	0.363
		min	0.008	0.014	0.089	0.226	0.007	0.04	0.047
Sway RAO (m/m)	0°	max	0.016	0.256	0.33	0.256	0.286	0.274	0.269
		min	0.005	0.028	0.133	0.184	0.17	0.194	0.12
	30°	max	0.786	0.903	0.69	0.804	0.824	0.737	1.011
		min	0.103	0.111	0.115	0.247	0.276	0.177	0.133
	45°	max	1.125	1.344	0.949	1.088	1.102	1.096	1.307
		min	0.188	0.135	0.138	0.213	0.319	0.171	0.121
	60°	max	1.434	1.659	1.218	1.306	1.356	1.397	1.504
		min	0.247	0.141	0.236	0.211	0.394	0.284	0.117
	90°	max	1.611	1.901	1.565	1.528	1.578	1.525	1.576
		min	0.013	0.246	0.202	0.27	0.524	0.262	0.194
Heave RAO(m/m)	0°	max	0.43	0.412	0.563	0.547	0.406	0.529	0.534
		min	0.042	0.017	0.123	0.122	0.013	0.089	0.094
	30°	max	0.467	0.443	0.494	0.666	0.504	0.427	0.691
		min	0.008	0.074	0.057	0.114	0.043	0.078	0.1
	45°	max	0.457	0.415	0.415	0.641	0.519	0.424	0.576
		min	0.018	0.128	0.021	0.122	0.123	0.067	0.093
	60°	max	0.452	0.556	0.43	0.6	0.633	0.439	0.557
		min	0.06	0.126	0.027	0.129	0.133	0.038	0.132
	90°	max	0.463	0.655	0.429	0.481	0.705	0.495	0.416
		min	0.042	0.162	0.114	0.08	0.147	0.079	0.098

(*Continued*)

Table 2.4 (Continued)

Description			Deck	BLS 1	BLS 2	BLS 3	BLS 4	BLS 5	BLS 6
Roll RAO (deg/m)	0°	max	0.007	0.122	0.161	0.12	0.126	0.136	0.127
		min	0.003	0.04	0.072	0.093	0.07	0.096	0.057
	30°	max	0.126	0.156	0.19	0.206	0.162	0.194	0.19
		min	0.021	0.034	0.026	0.082	0.12	0.069	0.021
	45°	max	0.172	0.216	0.211	0.22	0.183	0.221	0.208
		min	0.037	0.007	0.041	0.091	0.125	0.06	0.065
	60°	max	0.21	0.282	0.247	0.218	0.26	0.257	0.245
		min	0.041	0.015	0.058	0.083	0.153	0.079	0.059
	90°	max	0.248	0.354	0.219	0.235	0.35	0.234	0.224
		min	0.015	0.041	0.053	0.092	0.188	0.09	0.057
Pitch RAO (deg/m)	0°	max	0.254	0.257	0.266	0.258	0.264	0.247	0.214
		min	0.039	0.042	0.082	0.073	0.048	0.125	0.102
	30°	max	0.24	0.24	0.218	0.316	0.24	0.262	0.285
		min	0.018	0.018	0.049	0.054	0.018	0.139	0.129
	45°	max	0.196	0.196	0.179	0.291	0.196	0.244	0.251
		min	0.017	0.017	0.081	0.052	0.017	0.142	0.131
	60°	max	0.14	0.141	0.199	0.252	0.143	0.223	0.201
		min	0.028	0.028	0.057	0.046	0.028	0.105	0.128
	90°	max	0.108	0.028	0.163	0.16	0.02	0.185	0.159
		min	0.004	0.006	0.029	0.005	0.004	0.108	0.118
Yaw RAO (deg/m)	0°	max	0.019	0.017	0.038	0.031	0.017	0.046	0.031
		min	0.004	0.007	0.014	0.01	0.008	0.021	0.015
	30°	max	0.022	0.022	0.056	0.063	0.022	0.068	0.045
		min	0.006	0.003	0.004	0.014	0.003	0.001	0.016
	45°	max	0.019	0.017	0.055	0.06	0.017	0.052	0.049
		min	0.003	0.008	0.009	0.018	0.006	0.004	0.021
	60°	max	0.013	0.013	0.051	0.053	0.013	0.059	0.055
		min	0.005	0.005	0.014	0.023	0.005	0.01	0.021
	90°	max	0.005	0.004	0.052	0.057	0.005	0.061	0.073
		min	0.003	0.006	0.009	0.021	0.007	0.018	0.022

Table 2.4 (Continued)

Description			Deck	BLS 1	BLS 2	BLS 3	BLS 4	BLS 5	BLS 6
Tether tension (MN)	0°	max	—	91.70	94.80	96.20	92.40	94.80	95.40
		min	—	91.00	93.70	94.50	91.30	94.10	94.00
		Maximum tether tension variation (%) = 13.17							
	30°	max	—	91.90	94.40	96.10	92.60	94.70	95.30
		min	—	91.00	93.60	94.50	91.50	94.10	94.00
		Maximum tether tension variation (%) = 13.05%							
	45°	max	—	92.10	94.20	95.90	92.90	94.60	95.20
		min	—	91.00	93.60	94.70	91.60	94.10	94.00
		Maximum tether tension variation (%) = 12.82%							
	60°	max	—	92.30	94.30	95.40	93.10	95.10	94.60
		min	—	91.00	93.70	94.40	91.60	94.20	94.10
		Maximum tether tension variation (%) = 12.23%							
	90°	max	—	91.60	95.30	96.00	91.90	95.30	95.20
		min	—	91.00	93.90	94.60	91.30	94.10	94.00
		Maximum tether tension variation (%) = 12.94%							

responses of the deck are also less by 23 and 30%, respectively. It is also seen that the maximum variation in tether tension is about 12.33%, while the permissible limit is about 22% (API RP 2SK 2005).

2.6 Stability Analysis of the BLSRP

As discussed in the previous sections, buoyant legs connected to the seabed using a taut-moored system with high initial pre-tension enable rigid body motion in the vertical plane. Environmental loads induce dynamic tether tension variations, which in turn affect the stability of the platform. The BLSRP is now examined with different postulated failure cases, which are created by placing eccentric loads at different locations on the deck. This can result in dynamic tether tension variation, causing chaotic tension variations. A detailed numerical analysis is carried out for the BLSRP using the Mathieu equation of stability. Increasing the magnitude of the eccentric load and changing its position is expected to influence the fatigue life of the tethers. As the platform is positive-buoyant, high initial pre-tension on the tethers is necessary to ensure position restraint (Chandrasekaran et al. 2010). The Mathieu equation is often used to determine the stability of ships and offshore compliant platforms (Rho et al. 2002, 2003; Simos and Pesce 1997). The Mathieu equation is a special form of the Hill equation with only one harmonic mode, whose canonical form is given by:

$$\frac{d^2 f}{d\tau^2} + \left(\delta - q\cos(2\tau)\right)f = 0 \tag{2.6}$$

where δ and q are problem-specific Mathieu parameters (Chandrasekaran et al. 2006a, b; Patel and Park 1991). The stability equation for tethers, using a linear cable model, is effective in examining dynamic tether tension variations. The Mathieu equation used in the present study is derived from the idealized linear cable equation and is given by:

$$\delta_n = \frac{4}{M\omega^2}\left\{mg\frac{(I_2 + I_4)}{I_1}\right\} - (P + mgL)\frac{I_3}{I_1} \tag{2.7a}$$

$$q_n = \frac{2A}{M\omega^2}\frac{I_3}{I_1} \tag{2.7b}$$

where M is the total tether mass, ω is the wave frequency, m is the mass per unit length of the tether, g is acceleration due to gravity, P is the initial pre-tension, and A is the tension amplitude. The corresponding integrals of these equations are given by:

$$I_1 = \int_0^L X_n^2(x)dx \tag{2.8a}$$

$$I_2 = \int_0^L \frac{dX_n}{dx}dx \tag{2.8b}$$

$$I_3 = \int_0^L \frac{d^2 X_n}{dx^2} X_n dx \tag{2.8c}$$

$$I_4 = \int_0^L \frac{d^2 X_n}{dx^2} X_n (xdx) \tag{2.8d}$$

where X_n is given by:

$$X_n(x) = J_0\left(2\beta_n\left[1 + \frac{mg(L-x)}{P}\right]^{1/2}\right) - \frac{J_0(2\beta n)}{Y_0(2\beta n)}Y_0\left(2\beta_n\left[1 + \frac{mg(L-x)}{P}\right]^{1/2}\right) \tag{2.9}$$

One of the Mathieu parameters, β_n, is obtained as the solution of the following equation:

$$J_0\left(2\beta_n\sqrt{1+\frac{mgL}{P}}\right)Y_0(2\beta_n)-Y_0\left(2\beta n\sqrt{1+\frac{mgL}{P}}\right)J_0(2\beta_n)=0 \tag{2.10}$$

Because the stability condition is influenced by the Mathieu parameters, the solution to the Mathieu equation is expressed in the form of a stability chart (Chandrasekaran 2015a, b, 2016a, b, 2017; Chandrasekaran and Jain 2016).

Each buoyant leg is modeled as a tubular member with outer diameters 14.14 and 0.15 m thick. A deck 100 m in diameter is connected to each buoyant leg using hinged joints, ensuring partial isolation of the deck; transfer of rotational responses from the legs to the deck is restrained. The buoyant legs are moored to the seabed using taut-moored tethers with high initial pre-tension. Each leg consists of a group of 4 tethers; a total of 24 tethers hold down the platform with a spread-mooring system. Figure 2.10 shows the numerical model in a moored condition, which is referred to as a *normal case* in the analysis.

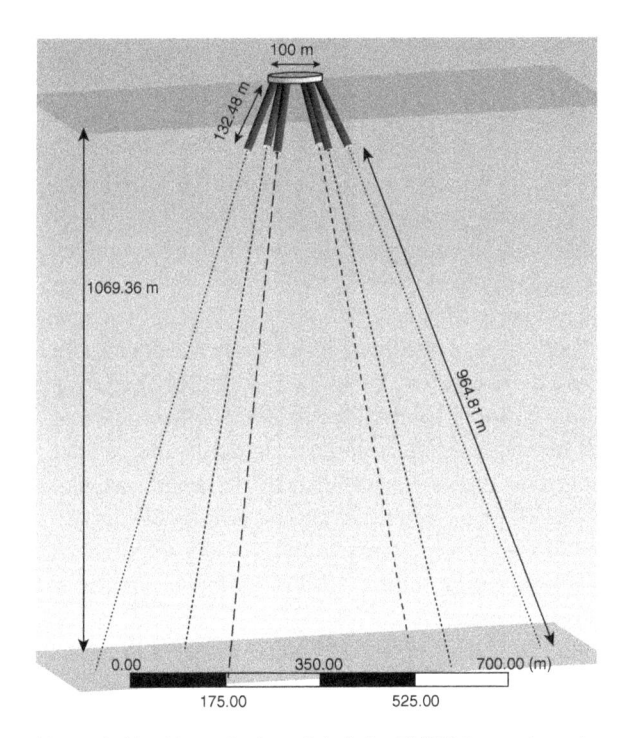

Figure 2.10 Numerical model of the BLSRP (normal case).

Table 2.5 Geometric properties of the BLSRP for the stability study.

Description	Triceratops
Water depth	1069.36 m
Total mass	641 000 kN
Buoyant force	940 880 kN
Diameter of the buoyant leg	14.14 m
Diameter of the deck	99.40 m
Length of the buoyant leg	132.48 m
Total tether force	319 125.60 kN
Pre-tension in each leg	53 187.61 kN
Tether length	964.81 m
Number of tethers (6 groups)	24
Axial stiffness of the tethers	76 830.67 kN/m

The static equilibrium between the buoyancy force, weight, and initial tether tension is as follows:

$$F_b = W + 6T_0 \cos(20) \tag{2.11}$$

It is important to note that a maximum vertical inclination of 20° is allowed for the legs. Table 2.5 shows the geometric properties of the platform. Three cases under eccentric loads are analyzed: (i) an eccentric load on top of one buoyant leg (referred to as case 2), (ii) an eccentric load between two adjacent buoyant legs (referred to as case 3), and (iii) an eccentric load on top of two adjacent buoyant legs (referred to as case 4). Such cases of eccentric loading are accidental and hence referred to as *postulated failure cases* in the study. Figure 2.11 shows the numerical model with different positions of the eccentric loads. Each case of eccentric loading is analyzed for two load magnitudes: 5 and 10% of the total mass. The BLSRP is analyzed given a regular wave (5 m, 6.8 s) in the normal and postulated failure cases. The BLSRP under eccentric loading is analyzed for two loads: 32 050 kN (5%) and 64 100 kN (10%). The maximum tension amplitude is summarized in Table 2.6, while dynamic tether tension variations for a few postulated failure cases are shown in Figure 2.12.

Postulated failure cases created by placing eccentric loads at different locations resulted in dynamic tether tension variations; chaotic tension variation is also observed in a few cases. Each case is examined for Mathieu instability. Using the equations discussed previously, the Mathieu parameters are obtained and plotted

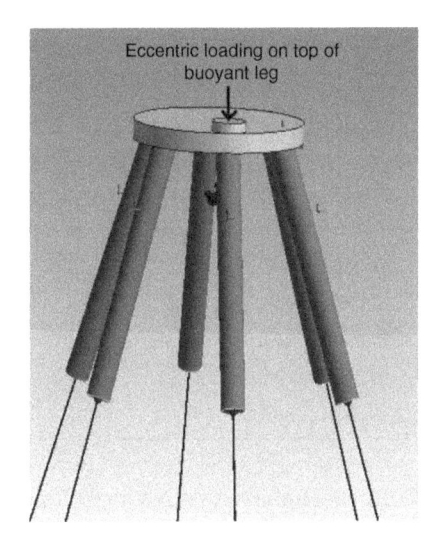

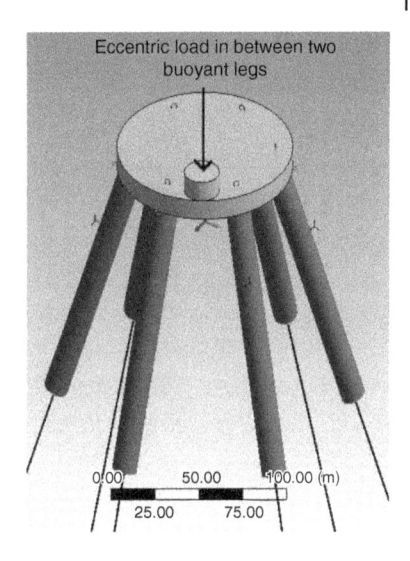

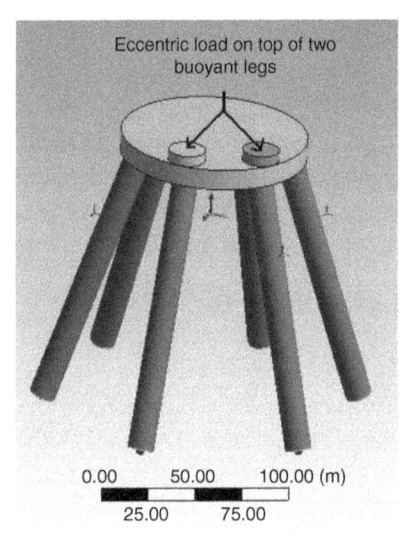

Figure 2.11 Numerical model of the BLSRP with postulated failure.

in the extended Mathieu stability chart, as shown in Figure 2.13. A summary of the results is shown in Table 2.7. As observed from the table, one of the parameters (δ), which depends on the stiffness and initial pre-tension of the tethers, remains constant for all the postulated failure cases. Another parameter (q), which depends on tension variation, differs for various postulated failures. It is also seen from the table that the platform is stable given normal conditions (case 1). Even eccentric

Table 2.6 Maximum tension amplitude in the tethers in postulated failure cases.

Description	Load	Leg 1 (MN)	Leg 2 (MN)	Leg 3 (MN)	Leg 4 (MN)	Leg 5 (MN)	Leg 6 (MN)	Maximum (MN)
Case 1	–	62.49	61.53	61.43	61.40	61.85	60.97	62.49
Case 2	5%	89.99	71.78	59.08	63.56	65.76	73.70	89.99
	10%`	168.54	140.70	76.40	128.80	80.79	136.04	168.54
Case 3	5%	85.06	18.52	68.28	67.98	65.11	77.17	85.06
	10%`	153.42	112.99	103.06	100.50	110.26	144.23	153.42
Case 4	5%	82.34	64.45	64.91	64.57	63.22	69.21	82.34
	10%`	112.36	99.39	73.37	68.76	80.14	107.94	112.36

loads in various postulated failure cases with 5% load amplitude did not result in Mathieu instability. For eccentric loading with 10% load amplitude, cases 2 and 3 show unstable conditions, justifying the chaotic nature of tether tension variation. It is interesting to note that given eccentric loads with a magnitude of 10% of the mass of the platform placed on the adjacent buoyant legs (case 4), the platform is in a stable condition. This is because the amplitude of the tension variation, which resulted in chaotic variation at first, settles down to a lower amplitude. Irrespective of the position of the eccentric load, the platform undergoes Mathieu instability for an eccentric load greater than 10% of the total mass of the structure.

2.7 Fatigue Analysis of the BLSRP

Tethers in taut-moored compliant structures are subjected to cyclic loading throughout their life. As seen previously, dynamic tether tension variation is significant in the postulated failure cases. Even though the amplitude of tension variation is less than the tether-breaking load, cyclic loading may lead to fatigue failure. The current study also investigates the fatigue life of tethers using the Miner-Palmgren approach. Fatigue strength is estimated based on the number of cycles (for example, 10^7) during which the maximum stress range can be applied without causing failure. The S–N curve is defined by the following equation:

$$A = NS^m \tag{2.12}$$

where S is the cyclic stress range, N is the number of cycles to failure, and A and m are constants depending on the fatigue class and the number of cycles. While the stress range and number of cycles are estimated using the rainflow-counting method, Miner's hypothesis is used to obtain the fractional damage caused by

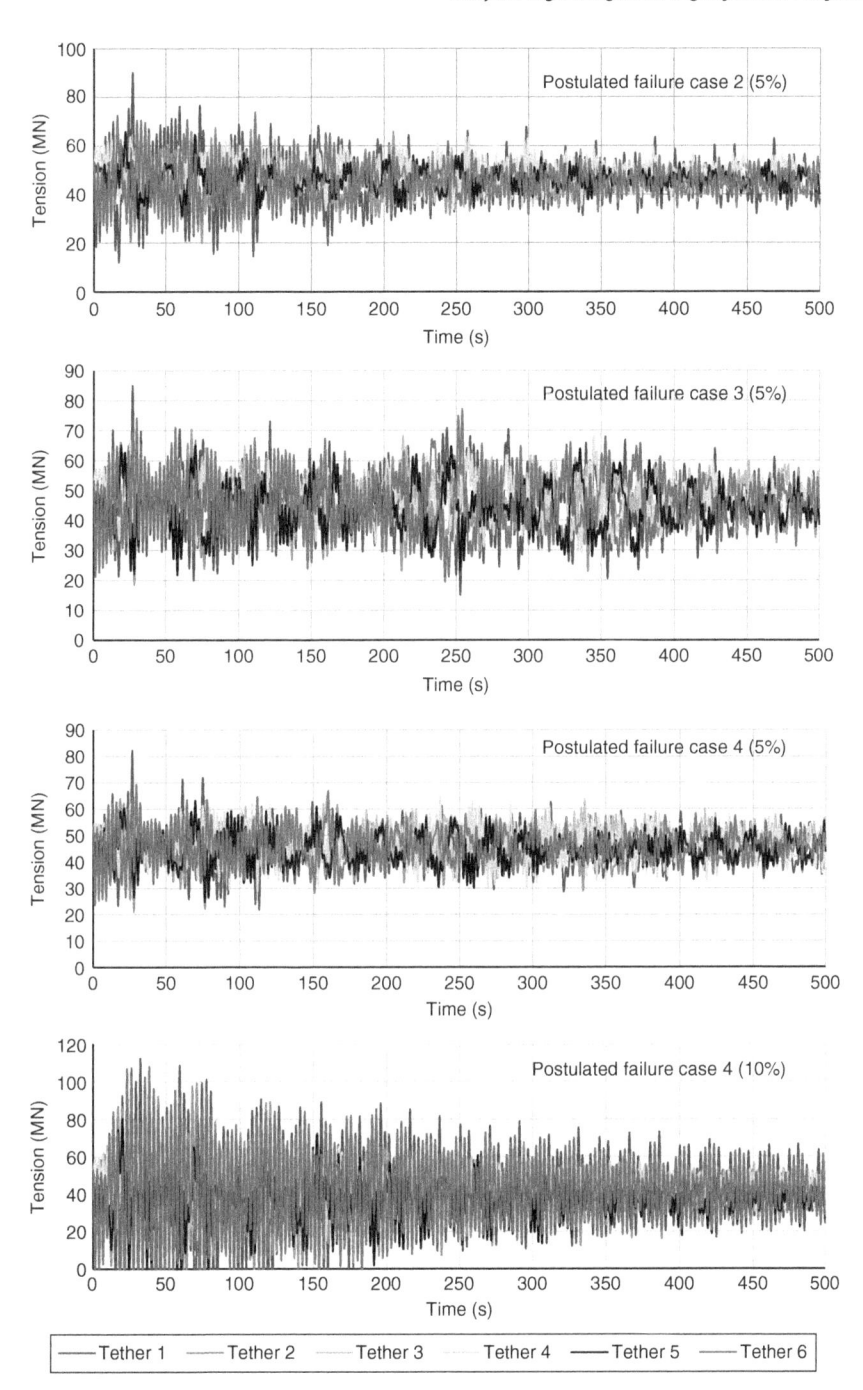

Buoyant Leg Storage and Regasification Platforms | **91**

Figure 2.12 Dynamic tether tension variation in postulated failure cases.

92 | Offshore Compliant Platforms

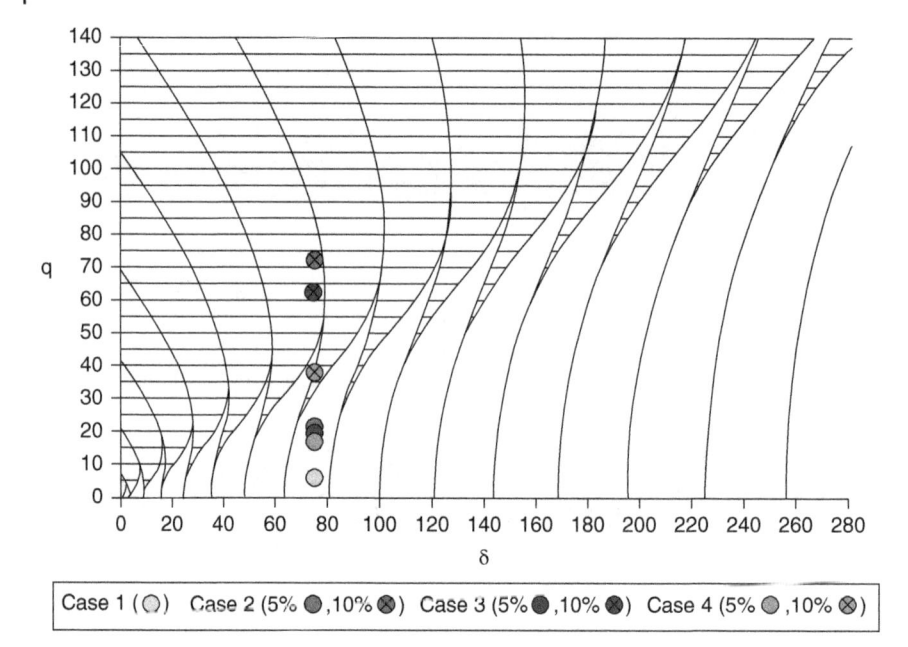

Figure 2.13 Mathieu stability for the BLSRP in postulated failure cases.

Table 2.7 Mathieu parameters in postulated failure cases.

Description	Load	δ	q	Stability condition
Case 1	—	75.07	5.9	Stable
Case 2	5%	75.07	23.35	Stable
	10%	75.07	73.19	Unstable
Case 3	5%	75.07	20.22	Stable
	10%	75.07	63.6	Unstable
Case 4	5%	75.07	18.49	Stable
	10%	75.07	37.54	Stable (boundary)

different stress ranges; the results are then summed up to obtain the overall damage, based on which the life of the tether is extrapolated. Damage is given by the following relationship:

$$D = \sum_{i=1}^{m} \frac{n_i}{N_i} \tag{2.13}$$

Table 2.8 Fatigue life (rounded off) of tethers under eccentric loading.

Description	Load	Leg 1	Leg 2	Leg 3	Leg 4	Leg 5	Leg 6	Min life
Case 1	—	23 Y	33 Y	33 Y	23 Y	34 Y	34 Y	23 Y
Case 2	5%	2 Y	5 Y	23 Y	13 Y	20 Y	5 Y	2 Y
	10%	14 days	29 days	2 Y	92 days	1 Y	30 days	14 days
Case 3	5%	3 Y	7 Y	15 Y	11 Y	9 Y	4 Y	3 Y
	10%	13 days	45 days	127 days	122 days	51 days	14 days	13 days
Case 4	5%	5 Y	13 Y	14 Y	8 Y	13 Y	9 Y	5 Y
	10%	103 days	286 days	5 Y	4 Y	1 Y	113 days	103 days

where D is the total damage, m is the number of stress bins, n is the number of stress cycles, and N is the number of stress changes. Detailed fatigue analyses are carried out for each tether in the postulated failure cases to obtain the service life of the tethers. A summary of the results is shown in Table 2.8. In a normal case, a maximum of 34.25 years of life is obtained for the tethers, whereas a minimum of 23.15 years is noted for the tethers of the wave-entrant buoyant leg. For 5% eccentric loading, a maximum reduction of 89.9% in the fatigue life is observed. As seen in cases 2 and 3, for a 10% load, the fatigue life of the tethers is reduced significantly to about 13 days, which is quite alarming. An increase in the magnitude of eccentric loading and the position of the load are very important. There is a significant decrease in fatigue life with the increase in the amplitude of tension variation. The very low fatigue life of tethers with Mathieu instability proves the severity of the instability. For example, case 4 under 10% loading shows a stable condition, but the fatigue life is very low compared to other stable condition cases.

3

New-Generation Platforms

Offshore Triceratops

Summary

This chapter explains the dynamic response of an offshore triceratops in response to the action of regular and random waves. Detailed numerical studies carried out on a triceratops in ultra-deepwater conditions are presented. The behavior of the structure is also compared to the previous experimental and numerical studies carried out on the offshore triceratops. This chapter also explains the behavior of the triceratops in response to the combined action of wind, wave, and current. The response spectra given different degrees of freedom and sea conditions are explained. A detailed methodology for the fatigue analysis of tethers is also presented, with results.

3.1 Introduction

Waves play a crucial role in the design of offshore structures due to the complicated hydrodynamic behavior exhibited by such structures in open sea conditions. The wave loads developed on offshore structures are much higher than the loads developed due to wind action. Wave loads that develop as a result of water particle motion also depend upon the size of the structural members. Due to the lower influence on the wave field of slender members with a diameter to wavelength ratio (D/L) of less than 0.2, the wave loads in such cases are calculated using the Morison equation. For large-diameter structures that affect the wave field, wave loads are calculated using diffraction theory. Wave loads are the most important of all environmental loads, so a detailed dynamic response analysis of any offshore platform is important in order to assess its suitability, especially in the case of

---

Offshore Compliant Platforms: Analysis, Design, and Experimental Studies,
First Edition. Srinivasan Chandrasekaran and R. Nagavinothini.
© 2020 John Wiley & Sons Ltd.
This Work is a co-publication between John Wiley & Sons Ltd and ASME Press.

ultra-deepwater conditions. The analysis of structures under wave loads greatly depends upon the type of waves considered.

Offshore structures are also subjected to other environmental loads arising due to the action of wind and currents, in addition to waves. Due to the fluctuating wind component, compliant offshore structures are more susceptible to low-frequency oscillations. The response of the platform also increases with an increase in the exposed area of the structure and wind velocity. In addition to wind effects, the current action causes varying pressure distribution around the offshore structure, resulting in steady drag force. Because the triceratops is a hybrid system with natural frequencies in two different bands, the combined actions of wind, wave, and currents results in complex response behaviors. The analysis in this case is necessary to understand the behavior of the structure in real sea conditions.

3.2 Environmental Loads

3.2.1 Regular Waves

Ocean waves are generally random. They can also be represented as regular waves and described using several wave theories. In a regular wave, the form of each cycle is the same; the wave theories describe the characteristics of one cycle based on major parameters such as wave period (T) and wave height (H). A typical time history of a regular wave is shown in Figure 3.1. The wave force acting on the

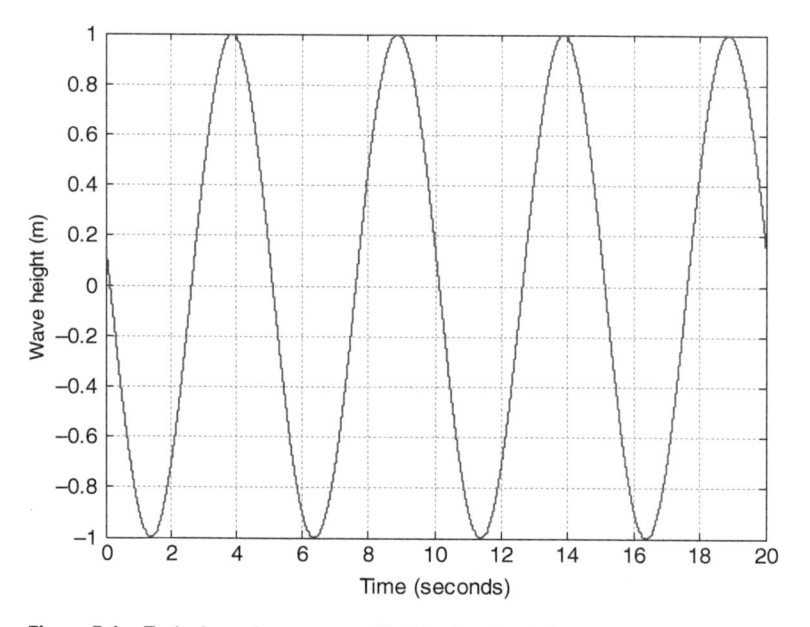

Figure 3.1 Typical regular wave profile (H = 2 m, T = 5 s).

structure can be calculated based on wave theories by obtaining the water particle kinematics as a function of the water surface elevation. The theories usually assume the waves are long-crested, and several wave theories have been developed for a wide range of wave parameters. The simplest wave theory used to evaluate the dynamic response of structures is Airy wave theory or small-amplitude wave theory. The wave is considered to have a sinusoidal profile, and the theory provides the dynamic and kinematic amplitudes as a linear function of wave height or wave amplitude. Thus, the resulting normalized amplitude will be unique and does not vary with respect to wave amplitude. So, the response of the structures under consideration can also be expressed as a normalized value. This normalized value of the structural response as a function of wave height is called the response amplitude operator (RAO) or transfer function. This method is simple and is commonly adopted to predict the response of structures even in extreme sea conditions.

3.2.2 Random Waves

In reality, ocean waves are a combination of waves with different frequencies and directions and appear to be irregular or random. Hydrodynamic analysis of structures in the presence of random waves is necessary to find the exact response of the structure in real sea conditions. Random waves are usually represented by wave energy density spectra that describe the spread of the wave's energy content from zero to infinite frequencies. Random waves are usually described using statistical parameters such as the significant wave height (H_s) and zero-crossing periods (T_z). An example of the classification of different sea conditions based on statistical parameters is given in Table 3.1. The most probable wave force is then calculated using linear wave theory. This statistical approach helps in more accurately assessing the dynamic behavior and fatigue strength of the structure by yielding the response spectrum, which clearly defines the maximum response of the structure in a particular interval of time.

Several wave spectrum models have been developed for the analysis and design of offshore structures in the presence of random wave action. These empirical models are derived based on the observed ocean properties, and the frequency characteristics of the real sea conditions significantly affect the spectral formulation

Table 3.1 Characteristics of random sea conditions.

Sea condition description	Significant wave height (m)	Zero-crossing period (s)
Moderate	6.5	8.15
High	10	10
Very high	15	15

(Chandrasekaran 2015a, b). The widely used wave spectra in offshore engineering are as follows:

- Pierson-Moscowitz (PM) spectrum
- Joint North Sea Wave Project (JONSWAP) spectrum
- International Ship Structures Congress (ISSC) spectrum
- Bredneidger spectrum
- Ochi-Hubble spectrum

The wave energy distribution across the frequency band differs with respect to the spectral model under consideration. Thus, the response of the structure will also vary for the same wave height based on the wave spectrum used in the analysis. The most commonly used wave spectrum in offshore design is the PM spectrum, which applies to different regions such as the Gulf of Mexico, offshore Brazil, western Australia, offshore Newfoundland, and western Africa, both in operational and survival conditions. This spectrum is suitable for representing open sea conditions, which are neither fetch limited nor duration limited. It is given by the following relationship:

$$S^+(\omega) = \frac{1}{2\pi} \frac{H_s^2}{4\pi T_z^2} \left(\frac{2\pi}{\omega}\right)^2 \exp\left(-\frac{1}{\pi T_z^4}\left(\frac{2\pi}{\omega}\right)^4\right)$$

where H_s is the significant wave height, T_z is the zero-crossing period, and ω is the frequency. Typical PM spectra for different sea conditions are shown in Figure 3.2.

The random wave time history for the time response analysis of structures can be developed from the wave spectrum by using inverse fast Fourier transform (IFFT). A sample of the random wave time history is shown in Figure 3.3.

3.2.3 Wind

A response analysis of an offshore platform to waves alone is not realistic, as the major source of wave generation is wind. In addition to the wind-induced wave force, wind also generates load on the superstructure. The dynamic wind effect is significant in the case of offshore compliant structures like the triceratops, and hence the analysis should be carried out by considering both the mean wind component and the gust component. Wind velocity is the major parameter in the analysis of structures under wind loads, based on which the wind pressure developed on the structure can be calculated. The average wind velocity occurring over a one-hour period is taken as the steady wind velocity, which is typically measured at 10.0 m above mean sea level (MSL). The wind load acting on the deck of the platform will induce additional moment, resulting in an excessive pitch response.

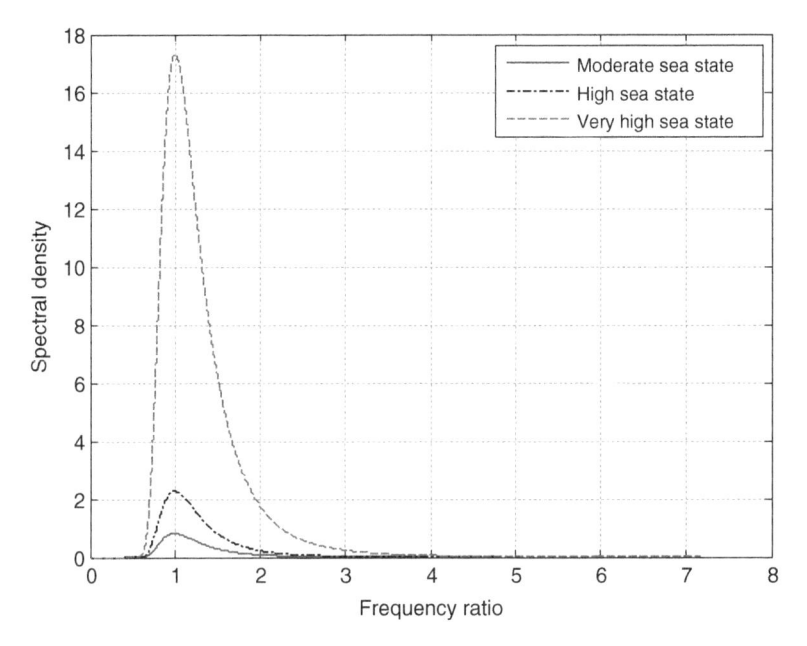

Figure 3.2 PM spectrum for different sea conditions.

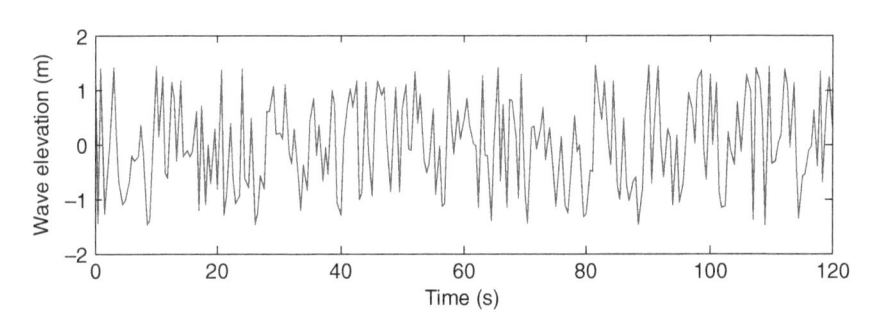

Figure 3.3 Two-dimensional random wave profile (Wang and Isberg 2015).

The combined action of wind and wave loads will result in deck offset and set-down as well, due to coupling between the surge and heave degrees of freedom (DOF) in compliant platforms. In addition, the additional moment induced by the wind load will result in coupling between the surge and pitch DOF.

Similar to the representation of random waves, the random wind blowing over a structure can be described using a wind spectrum. The wind spectra used to represent random wind in the analysis of structures are as follows:

- Davenport spectrum
- Harris spectrum

- Kaimal spectrum
- Simiu spectrum
- Kareem spectrum
- American Petroleum Institute (API) spectrum

These spectral formulations provide variations of wave spectral energy over a wide range of frequencies. The wind spectra exhibit significant differences at lower frequencies. Lower spectral energy occurs in the case of the Davenport spectrum. As both the Davenport and Harris spectra are developed for land-based conditions, they are not suitable for the analysis of offshore structures. One of the commonly used spectra for the analysis of offshore structures is the API spectrum, which shows higher energy at lower frequencies compared to other spectra. It is represented as follows:

$$\frac{\omega S_u^+(\omega)}{\sigma_u(z)^2} = \frac{\theta}{(1+1.5\theta)^{5/3}}$$

where θ is the frequency ratio or derivable variable $\left[\theta = \dfrac{\omega}{\omega_p}\right]$, ω_p is the peak frequency, z_s is the surface height (20 m), $\sigma_u(z)^2$ is the variance of U(t) at a reference height, z is reference height (=10 m), and $S_u^+(\omega)$ is the spectral density (Chandrasekaran 2015a, b). The variance $\sigma_u(z)^2$ at the reference height is given by:

$$0.01 \leq \frac{\omega_p^2}{U_z} \leq 0.1$$

$$\sigma_u(z) = \begin{cases} 0.15\bar{U}_z \left(\dfrac{z_s}{z}\right)^{0.125} & (if \to 2 \leq z_s) \\ 0.15\bar{U}_z \left(\dfrac{z_s}{z}\right)^{0.275} & (if \to 2 > z_s) \end{cases}$$

The API spectra for different wind velocities are shown in Figure 3.4. As seen from the plot, the spectral energy is at a maximum in the lower-frequency region. Compared to wave spectra, which are narrow banded and have the maximum energy concentrated close to the wave frequency, the wind spectra are wide banded without any significant peaks.

3.2.4 Currents

Current generation in the sea is mainly due to the following factors:

- Wind effects
- Tidal motion

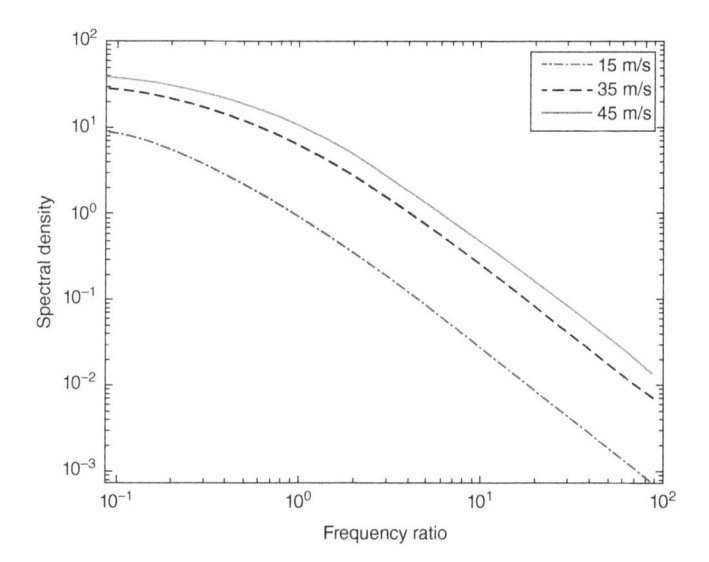

Figure 3.4 API spectrum plot for different wind velocities.

- Temperature differences
- Density gradients
- Salinity variations

The apparent wave period and the total water particle velocity are altered by the presence of currents. The current action also imposes additional drag forces on structures, which in turn affect the tether tension variation of compliant structures. Wind-generated currents are highly concentrated close to the sea surface, and the effect decreases with greater water depth. The current effect is included in the analysis by representing the current velocity, which varies linearly from a maximum value at the sea surface to zero at the seabed. The maximum velocity of wind-generated current can be approximated as 1.0–3.0% of the sustained wind velocity (Reddy and Swamidas 2016). The wind-generated current velocity profile in the Gulf of Mexico is shown in Figure 3.5. The current in the same direction as waves increases the wavelength and the wave period. The increased wave period (10%) due to current action is called the *apparent wave period*.

3.3 Fatigue Analysis of Tethers

Tethers are crucial components of compliant offshore platforms because the failure of one or more tethers may endanger the stability of the platform. It may also affect the functioning of the platform, which in turn affects oil production and return on investment. Tether analysis of offshore structures is mandatory

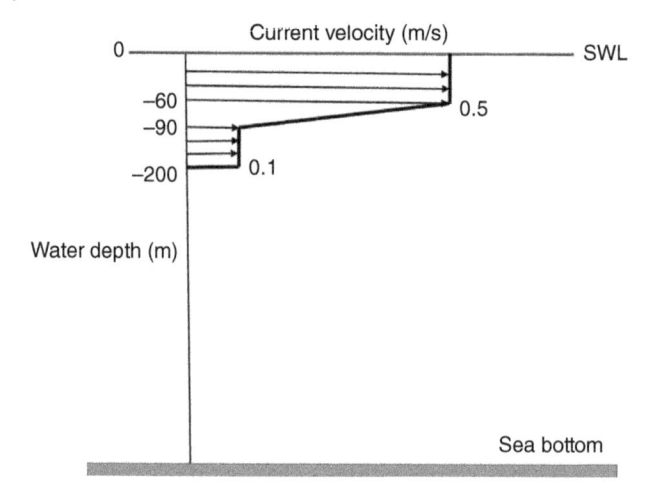

Figure 3.5 Wind-generated current velocity profile.

from both the structural and economical points of view. In the case of a triceratops, a set of tethers is used to connect the three buoyant legs to the seabed. These tethers are under very high initial tension, similar to the tethers of tension leg platforms. The action of loads on the structure induces a change in the initial tension of the tethers. Tether tension variation of the triceratops increases with greater severity of the sea conditions. However, the periodic tether response may impose fatigue in the tethers. Thus, the service life of the tethers can be predicted through fatigue analysis using the methodology shown in Figure 3.6.

The steps involved in the tether fatigue analysis are as follows:

Step 1: Dynamic response analysis of the triceratops
   The dynamic response analysis of the triceratops should be carried out under the action of either environmental loads or accidental loads through experimental or numerical investigations.

Step 2: Tether tension variation
   The tension variation of a tether should be obtained from the investigations carried out on the structure.

Step 3: Tether stress time history
   From the known area of the tether and the tether tension variation, the tether stress variation time history is obtained.

Step 4: Stress histogram
   The stress histogram should be developed from the stress time history. This stress histogram gives the stress range with the number of cycles.

```
┌─────────────────────────────────────────┐
│    Dynamic analysis of triceratops      │
└─────────────────────────────────────────┘
                    ↓
┌─────────────────────────────────────────┐
│         Tether tension variation        │
└─────────────────────────────────────────┘
                    ↓
┌─────────────────────────────────────────┐
│         Tether stress time history      │
└─────────────────────────────────────────┘
                    ↓
┌─────────────────────────────────────────┐
│            Stress histogram             │
│     (Using rainflow-counting method)    │
└─────────────────────────────────────────┘
                    ↓
┌─────────────────────────────────────────┐
│          Allowable stress cycles        │
│         (Using S–N relationship)        │
└─────────────────────────────────────────┘
                    ↓
┌─────────────────────────────────────────┐
│             Fatigue damage              │
│        (Using Miner-Palmgren rule)      │
└─────────────────────────────────────────┘
                    ↓
┌─────────────────────────────────────────┐
│          Service life estimation        │
└─────────────────────────────────────────┘
```

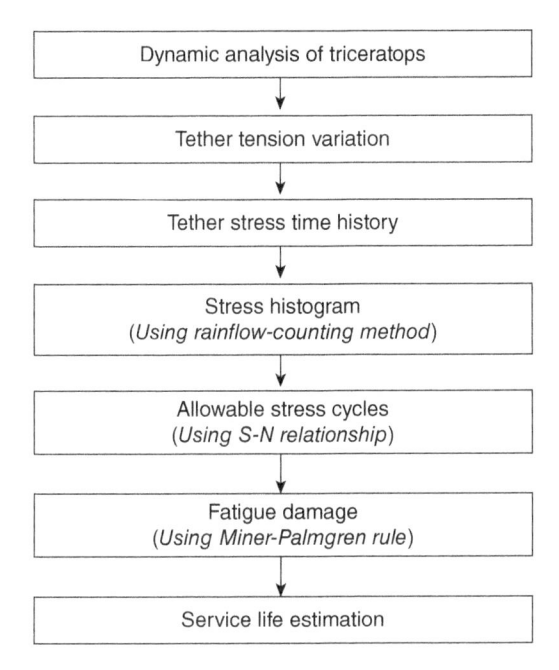

Figure 3.6 Service life estimation methodology.

Step 5: Allowable stress cycles

The allowable stress cycles should be calculated according to the standard regulations using an S–N curve approach (Veritas 2010a). It is given by:

$$\log N = \log B - m \log S$$

where N is the number of allowable cycles, S is the stress range, and B and m are constants obtained from the S–N curves.

Step 6: Fatigue damage assessment

The fatigue damage to the tether is then calculated using the Miner-Palmgren rule given by:

$$D_f = \sum_{i=1}^{m} \frac{n_i}{N_i}$$

where D_f is the fatigue damage, n is the number of stress counts from the histogram, and N is the number of allowable cycles from the S–N relationship.

Step 7: Service life calculation

Fatigue damage is then calculated for one year. Finally, the service life of the tethers is calculated by extrapolating the fatigue damage as unity.

3.4 Response to Regular Waves

Because the triceratops is a multi-legged structure with ball joints, it exhibits complex rigid-body motion under the action of lateral loads. The response also differs with respect to the direction of the wave action. The size of the structure and the water depth also affect the response of the platform significantly. The experimental investigation carried out on a 1:150 scale model of a triceratops, shown in Figure 3.7, showed less heave response compared to the surge response in the presence of regular wave action. The reduction in the rotational response of the deck provides a comfortable working environment for the crew onboard. In addition, structural integrity is ensured by the complete transfer of the heave response from the buoyant legs to the deck (Chandrasekaran et al. 2011). This basic study also includes a detailed investigation of the triceratops given different sea conditions. Suitable improvisations in the structural geometry are also attempted by investigating the behavior of the stiffened triceratops, which consists of buoyant legs made up of three cylindrical legs interconnected to a moon pool with stiffeners, as shown in Figure 3.8. The detailed experimental and numerical investigations of the triceratops at a water depth of 215.0 m given regular and random waves observed that the deck remains horizontal under the action of wave loads. This can be primarily attributed to the presence of the ball joints (Chandrasekaran and Madhuri 2015). In addition, the stiffening in the buoyant legs increases the tether tension significantly (Chandrasekaran and Mayank 2017).

Though the platform is found have several operational advantages, there is a need to assess the behavior of the same geometry in ultra-deepwater conditions, considering the recent increase in oil exploration at greater water depths. The

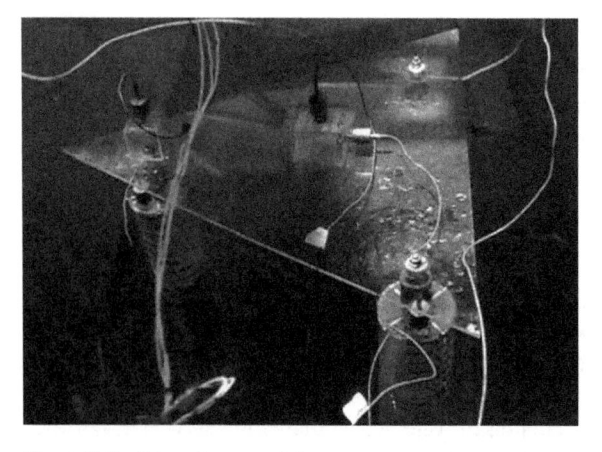

Figure 3.7 Triceratops model.

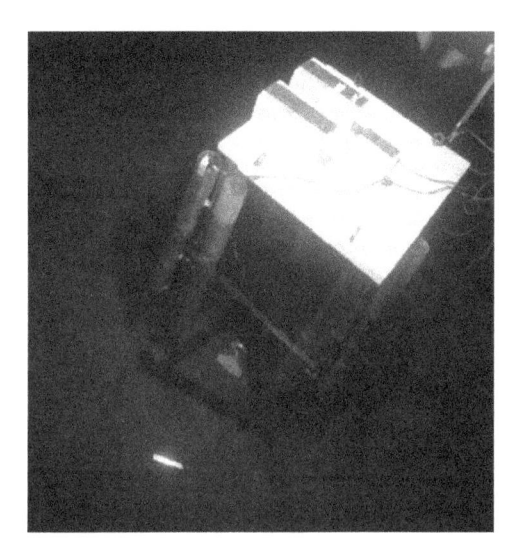

Figure 3.8 Experimental model of a stiffened triceratops.

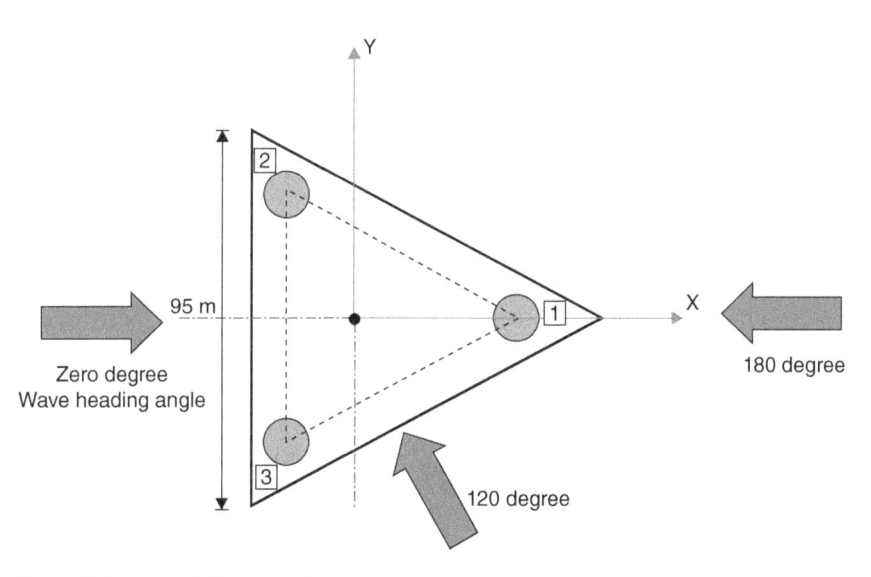

Figure 3.9 Plan of the triceratops.

behavior of the offshore triceratops developed for a water depth of 2400 m is assessed through a numerical investigation. Considering the three different wave heading angles as shown in Figure 3.9, wave action with 0° and 180° wave heading angles induces significant surge, heave, and pitch responses in both the deck and buoyant legs. The responses in the sway, roll, and yaw DOF are significant.

The responses of the deck and buoyant legs in the active DOF (surge, heave, and pitch) are periodic in nature, fluctuating around the mean position of the respective structural elements. The ball joints restrain the transfer of the pitch motion from the buoyant legs to the deck.

The RAO of the deck and buoyant legs given a 0° wave heading angle is shown in Figure 3.10. With an increase in the wave period, the surge response of both the buoyant legs and deck increases. However, coupling between the surge and pitch DOF decreases the surge response of the deck to less than that of the buoyant legs. This response behavior is considered to be one of the advantages of the adopted geometry, as the operations carried out on the topside are highly influenced by the surge response of the deck. In addition, the heave response of the deck is only about 1.0% of the surge response. Because the heave amplitude is small, it reduces the flexural yielding of the risers. The ball joints restrain the transfer of the pitch response from the buoyant legs to the deck by more than 90% (Chandrasekaran and Nagavinothini 2018a, b). The response of the offshore triceratops at 2400 m water depth and the stiffened triceratops at 215.0 m water depth are compared in Table 3.2. The reduced response of the triceratops in ultra-deep water shows the adequacy of the developed geometry.

The response of the triceratops varies with respect to the wave heading angles, which can be mainly attributed to the geometric configuration of the platform. Deck RAO plots at different wave heading angles are shown in Figure 3.11. In the case of a 120° wave heading angle, responses are observed in all DOF because the wave direction is not aligned with any of the global axes. The deck surge response given a 120° wave heading angle is lower than that of the response observed at 0° and 180° wave headings by 54% and 50% at a wave period of 15 seconds. In all cases, the maximum heave response is observed at the heave natural period. Though the 0° and 180° wave headings are aligned to the global x-axis, a significant variation is observed in the response of the platform. This is mainly due to the asymmetry of the platform with respect to the wave heading angles. At a 0° wave heading, two buoyant legs are located at the leading end, with the third leg at the trailing end; the reverse is true for a 180° wave heading angle. This results in differential heave in the buoyant legs, which also affects the surge response of the deck due to the coupling between the surge and heave DOF that is inherent in any compliant platform.

The platform is monolithic in the translational DOF, so the transfer of differential heave from the buoyant legs induces a small pitch response in the deck of the triceratops. However, the amplitude of the pitch response with different wave heading angles is much less compared to that of the buoyant legs, which makes the platform highly advantageous compared to other conventional offshore structures. Based on the studies, it can be said that aligning the global x-axis of the platform with the most prominent wave direction of the offshore site will result in reduced responses in different DOF.

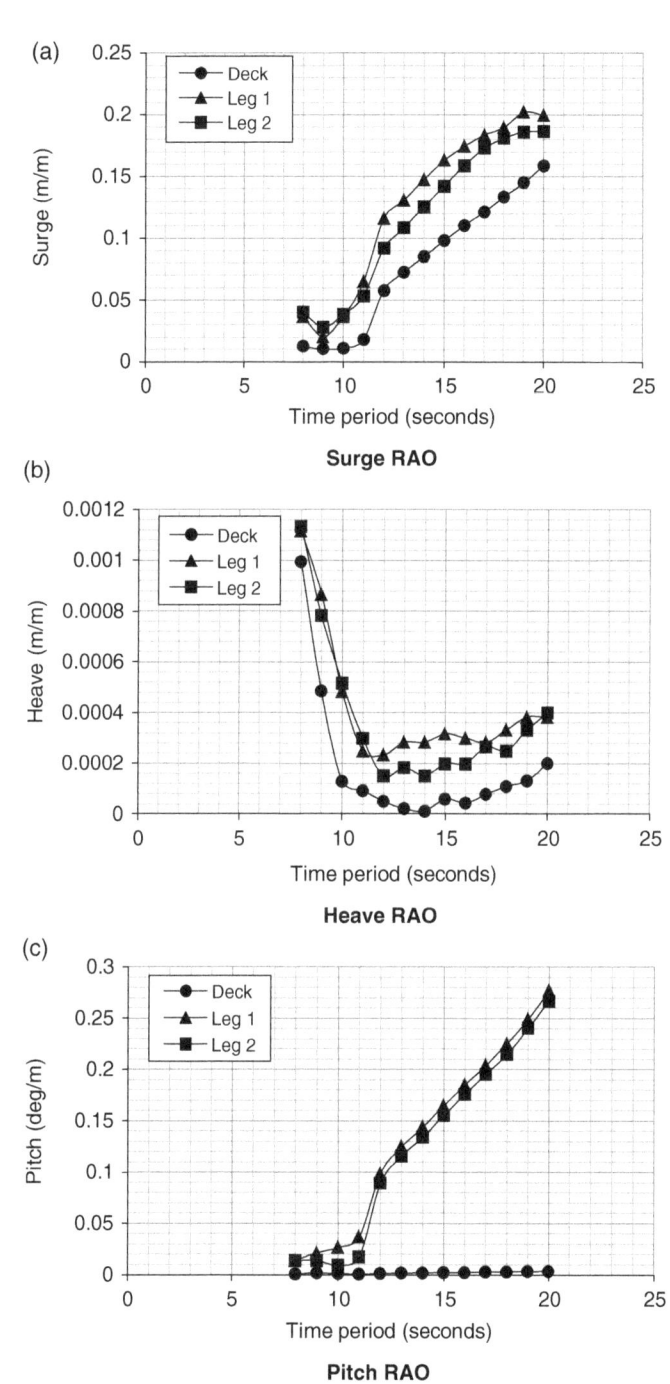

Figure 3.10 RAOs of the deck and buoyant legs with regular waves.

Table 3.2 Comparison of responses to regular waves.

Degrees of freedom	Triceratops (2400 m water depth)	Stiffened triceratops (215 m water depth)
Surge (m/m)	0.0979	0.38
Heave (m/m)	0.00006	0.0025
Pitch (deg/m)	0.0022	0.03

In the presence of the action of regular waves, the tether tension fluctuates around the initial tension value. The tether response is predominantly governed by the action of regular waves and the combined responses of the deck and buoyant legs of the triceratops. The variation in the response of the tethers of the different buoyant legs is marginal. Given a 0° wave heading angle, the maximum response occurs in the buoyant leg at the leading end. The tension variation of the tether that connects buoyant leg 1 to the seabed under rough sea conditions (wave height = 4.0 m, wave period = 9 seconds) is shown in Figure 3.12. The maximum tension developed under this sea condition is 3.60% higher than the initial tether tension. The maximum tether tension increases with increases severity of the sea conditions. Given phenomenal conditions (wave height = 24.0 m, wave period = 14 seconds), the tether tension increases by about 19% compared to the initial tension. The stress developed in the tether will not cause tether failure, but the periodic stress variation may impose fatigue damage on the tethers. The service life of the tethers calculated through fatigue analysis is about 20.03 years in the case of rough sea conditions. It decreases to 13.02 years under phenomenal sea conditions due to the variation in the maximum stress range and the number of stress cycles. Note that because the triceratops is a positively buoyant structure, tether failure will not result in the complete collapse of the structure. But the functionality of the platform will be affected due to the disruption caused to the risers by the tether failure.

3.5 Response to Random Waves

Given a 0° wave heading angle, the response of the triceratops in the active DOF increases given moderate to very high sea conditions. Studying the response statistics enables one to understand the response clearly. The response statistics of an offshore triceratops given different sea conditions at 2400 m water depth are given in Table 3.3. With an increase in the severity of the sea conditions, the surge mean shift also increases. Due to the coupling effect, a mean shift increase also occurs in

(a) Surge

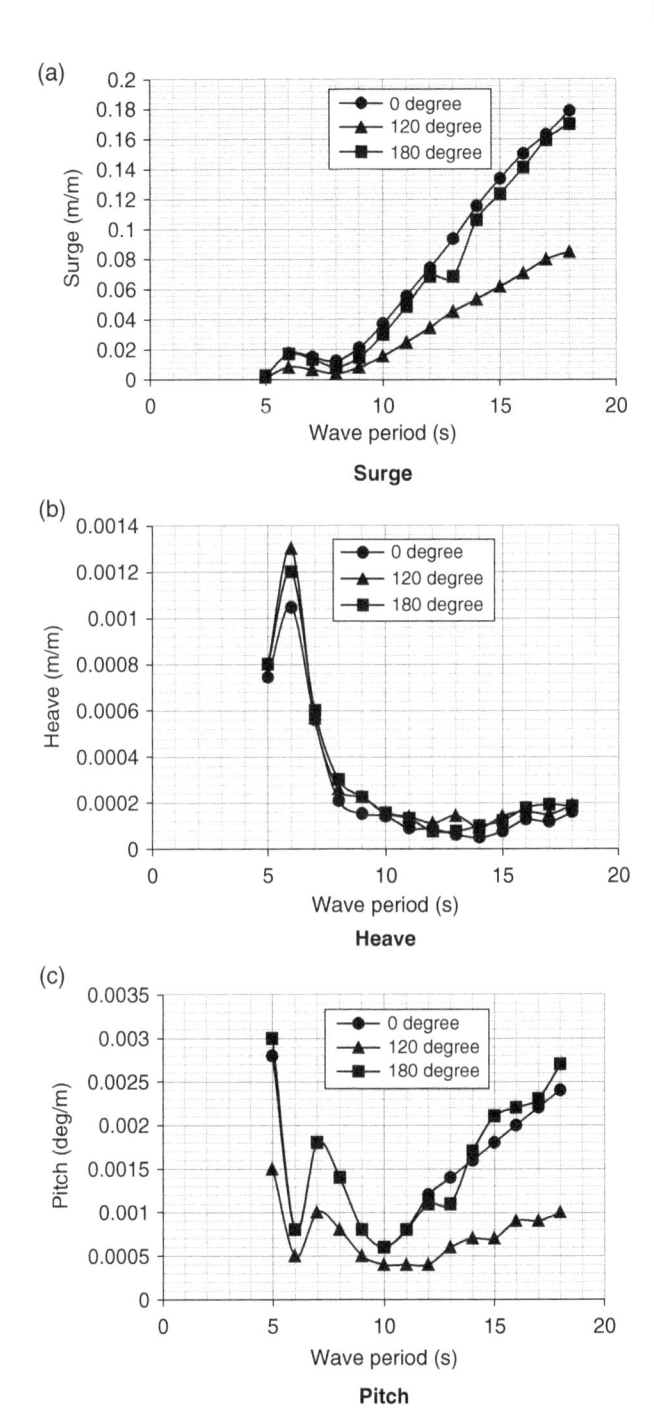

(b) Heave

(c) Pitch

Figure 3.11 Deck response given different wave heading angles.

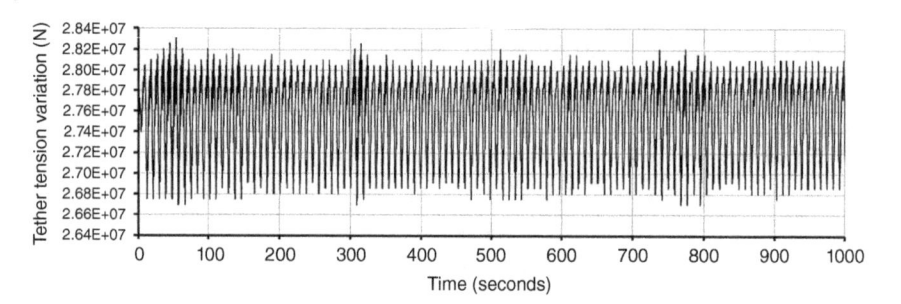

Figure 3.12 Tether tension variation in rough sea conditions.

Table 3.3 Deck response to different sea conditions.

Sea condition	Statistics	Surge (m)	Heave (m)	Pitch (deg)
Moderate	Maximum	2.301	0.005	0.030
	Minimum	−2.094	−0.015	−0.016
	Mean	0.030	−0.002	0.001
	Standard deviation	0.654	0.002	0.004
	RMS	0.655	0.003	0.004
High	Maximum	6.478	0.006	0.114
	Minimum	−5.034	−0.106	−0.049
	Mean	0.123	−0.008	0.002
	Standard deviation	1.654	0.012	0.010
	RMS	1.659	0.014	0.010
Very high	Maximum	23.911	0.007	0.582
	Minimum	−15.890	−2.245	−0.353
	Mean	1.239	−0.305	0.024
	Standard deviation	7.361	0.362	0.097
	RMS	7.465	0.474	0.100

the heave and pitch DOF. The increase in the response of the structure along with the roughness of the sea conditions is also reflected in the increased standard deviation of the responses in different DOF. The pitch response of the deck is much less even under very high sea conditions, showing the efficiency of ball joints in restraining the rotational motion from the buoyant legs to the deck.

In random wave analysis of structures, the dominant frequency at which the structure exhibits a maximum response can be studied through the response spectra. Power spectral density plots of deck surge and heave responses given very high sea conditions are shown in Figure 3.13. The peaks in the response spectrum usually occur at multiples or fractions of the structure's natural frequency or the dominant wave frequency. In this case, the maximum peak in the surge response spectrum occurs at one-fourth of the pitch natural frequency of 0.25 rad/s, and the first peak occurs at the surge natural frequency (0.029 rad/s). The maximum spectral energy is very close to one-fourth of the pitch natural frequency, which shows the coupling between the surge and pitch DOF (Chandrasekaran and Nagavinothini 2017).

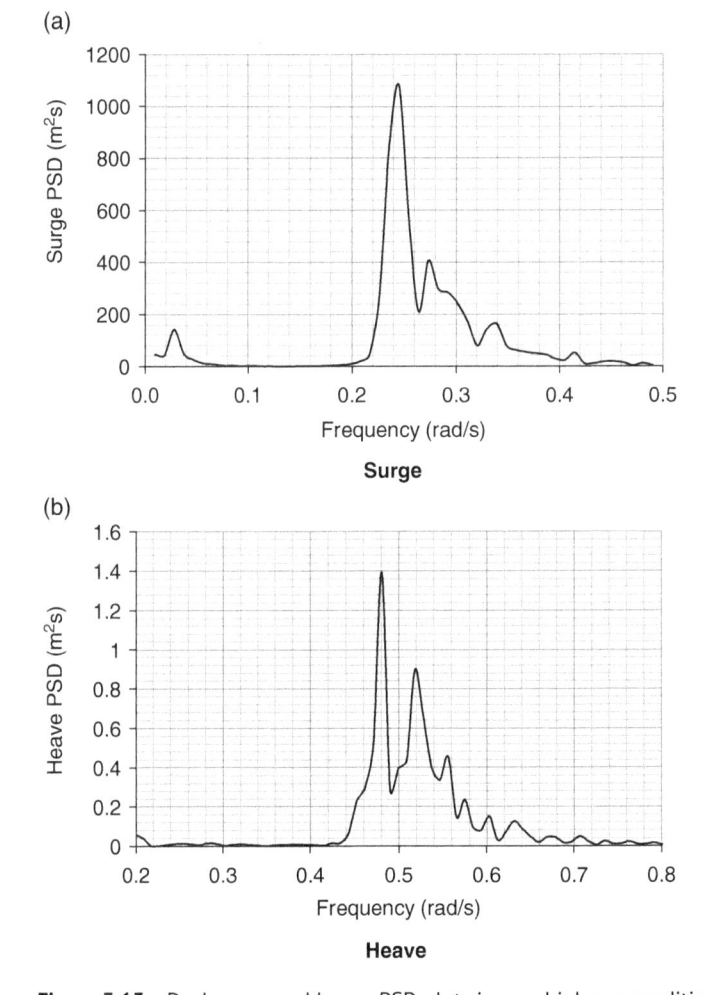

Figure 3.13 Deck surge and heave PSD plots in very high sea conditions.

Similar to the response to regular waves, the heave response of the deck is also less than the surge response to the action of random waves. The maximum peak in the heave response occurs at one-third of the heave natural frequency (0.48 rad/s). The pitch responses of the buoyant legs and deck of the triceratops given very high sea conditions are shown in Figure 3.14. The maximum peak in the pitch response of the buoyant legs occurs at one-fourth of the pitch natural frequency, whereas the peak in the deck response is 92.30% less than that of the buoyant legs. This is due to the restraint offered by the ball joints. However, complete restraint cannot be ensured due to the transfer of the unequal heave response from the three buoyant legs to the deck. A comparison of the response of the triceratops in ultra-deep water to the stiffened triceratops at 215.0 m water depth is shown in Table 3.4. The triceratops in ultra-deep water exhibits larger surge peak values, whereas the heave response remains the same in both cases. Despite the increase in the spectral peak value in the surge DOF, the maximum surge response is only about 1.08% of the water depth. The pitch peak value of the deck of the ultra-deepwater triceratops is 48% less than that of the stiffened triceratops.

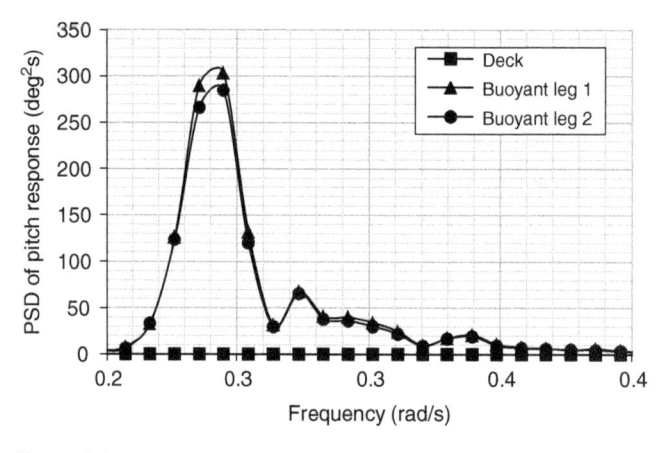

Figure 3.14 Pitch response of the deck and buoyant legs in very high sea conditions.

Table 3.4 Comparison of deck responses to high sea conditions.

Maximum deck response	Triceratops (2400 m water depth)	Stiffened triceratops (215 m water depth)
Surge PSD (m^2s)	26.00	16.00
Heave PSD (m^2s)	0.000 59	0.000 58
Pitch PSD (deg^2s)	0.000 042 4	0.000 088

The discrepancies in the results are mainly due to the difference in the geometric configuration of the buoyant legs in the studies.

In order to predict the complete hydrodynamic interaction of the platform, an assessment of the response of the triceratops to a range of wave approach angles is highly useful. The maximum deck responses in the surge, sway, and heave DOF given very high sea conditions are shown in Figure 3.15. The deck response in the rotational DOF is restrained by the ball joints. The maximum surge response decreases from 0° to 90° and then increases, whereas the sway response exhibits the opposite behavior. The variation in the heave response is marginal for the range of wave approach angles. The optimum wave approach angle that induces a reduced deck response in all translational DOF is 45°, where the responses in the surge and sway DOF match each other.

The tension variation in the tethers of different buoyant legs is marginal: on the order of 2.0%. Similar to the response of the triceratops in different DOF, the maximum tether tension and the mean shift from the initial tension increase with an increase in the roughness of the sea conditions, as seen from the tether tension statistics given for different sea conditions in Table 3.5. The tether tension power spectral density (PSD) given very high sea conditions is shown in Figure 3.16. With very high sea conditions, the energy band is close to the peak wave frequency, with a narrow peak at the pitch natural frequency. The service life of the tethers as calculated based on the fatigue analysis is less in the case of random wave action compared to regular wave action. The service life also decreases marginally with an increase in the roughness of the sea conditions. The failure of any one of the tethers of the buoyant legs may lead to successive failures of other tethers and endanger the stability of the whole platform. The wave approach angle significantly affects the tether tension variation in different buoyant legs, as seen in Figure 3.17. At a 60° wave approach angle, the maximum tether tensions in all the buoyant legs are closer. The tether tension in buoyant legs 1 and 3 is comparatively less than that of buoyant leg 2 at wave approach angles of 0° and 120°, respectively.

3.6 Response to Combined Actions of Wind, Waves, and Current

Because the ball joints in the triceratops restrain the transfer of rotational motion, they are also effective at restraining the transfer of yaw motion to the deck due to wind loads on the buoyant legs. This reduces the overall response of the structure (Chandrasekaran et al. 2013). The response of the structure varies with respect to the wave and wind parameters considered during the analysis. Considering the sea conditions for the analysis of an ultra-deepwater triceratops given the combined action of wind and wave loads, the response of the triceratops increases

114 | *Offshore Compliant Platforms*

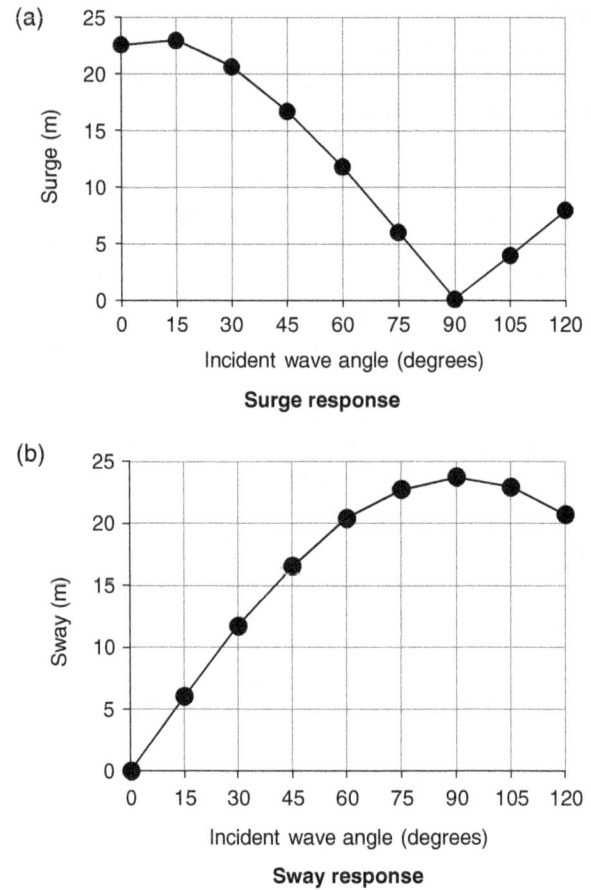

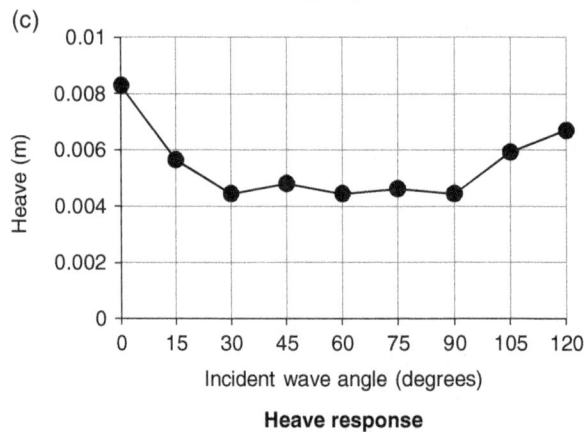

(a)

(b)

(c)

Figure 3.15 Maximum deck response in very high sea conditions.

New-Generation Platforms | 115

Table 3.5 Tension variation and service life of tethers of buoyant leg 1.

Statistics (MN)	Moderate	High	Very high
Maximum	29.474	29.958	30.388
Minimum	26.301	25.709	24.848
Mean	27.671	27.670	27.751
Tether tension variation (%)	5.73	7.68	9.98
Service life of tethers in years	14.02	14.01	13.83

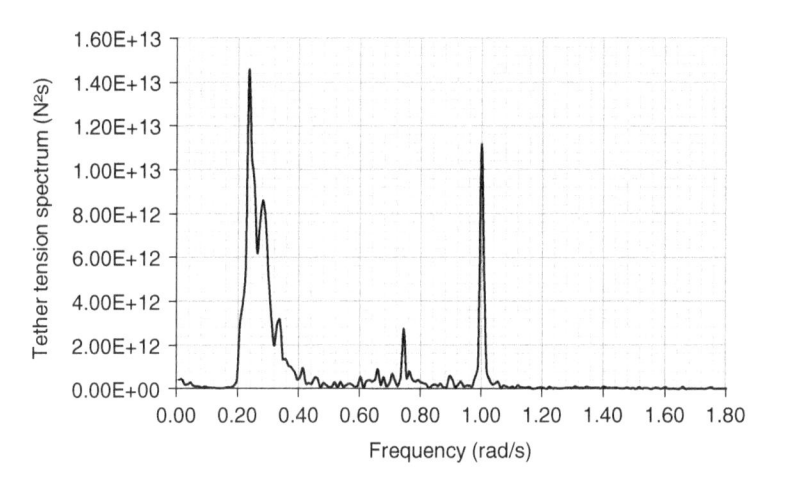

Figure 3.16 Tether tension spectrum with very high sea conditions.

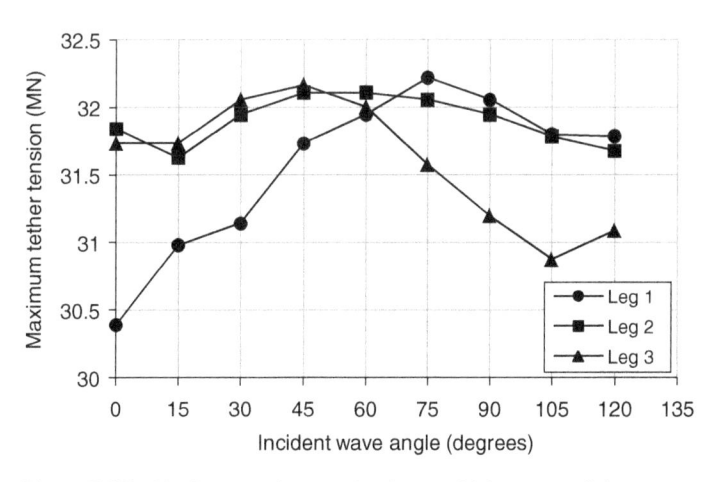

Figure 3.17 Maximum tether tension in very high sea conditions.

Table 3.6 Characteristics of sea conditions (Jain and Chandrasekaran 2004).

Sea condition description	Significant wave height, H_s (m)	Zero-crossing period, T_z (s)	Wind velocity (m/s)
Moderate	6.5	8.15	15
High	10	10	35
Very high	15	15	45

with increased the roughness of the sea conditions. The characteristics of sea conditions with wind velocity are given in Table 3.6.

3.6.1 Deck Response

The wind action increases the response of the deck in the active DOF (surge, heave, and pitch) at a 0° approach angle, compared to waves alone. The current also increases the maximum response of the deck along with the mean shift, which makes the structure oscillate at its new position. Despite the coupling between the surge and heave DOF, which is inherent in compliant platforms, the increases in the surge and heave responses of the deck due to the wind action are not equal. This is mainly due to the differential heave transfer from the buoyant legs to the deck. Coupling exists between the surge and pitch DOF due to the distance between the aerodynamic center and the mass center of the deck. This increases the pitch response of the deck to the combined wind and wave actions. It can be reduced by decreasing the distance between the aerodynamic center and mass center of the deck (Jain and Chandrasekaran 2004). The PSD plots of the response in the active DOF given high sea conditions are shown in Figure 3.18.

Because the triceratops has two sets of natural frequencies with respect to the stiff and flexible DOF, the energy in the response spectrum is concentrated close to multiples or fractions of the natural frequencies. In the surge response, the first peak occurs close to the surge natural frequency (0.028 rad/s) and the second peak occurs at one-fourth of the pitch natural frequency (0.226 rad/s). In comparison to an environment with waves alone, the wind action increases the spectral peak value by 2.96 and 7.52 times the surge natural frequency and one-fourth of the pitch natural frequency, respectively. Though the current action increases the maximum surge response with the mean shift, the total response of the deck is reduced due to the addition of current. The spectral energy in the surge response given the combined actions of wind, waves, and current is concentrated close to the surge natural frequency. The increased heave response due to the wind action can be seen from the increase in the spectral energy, close to half of the pitch natural frequency (0.45 rad/s). Similar to the surge response, the total heave response is also reduced by the addition of current. The magnitude of the deck pitch response under the combined action of different loads remains much less compared to that of the buoyant legs due to the presence of the ball joints.

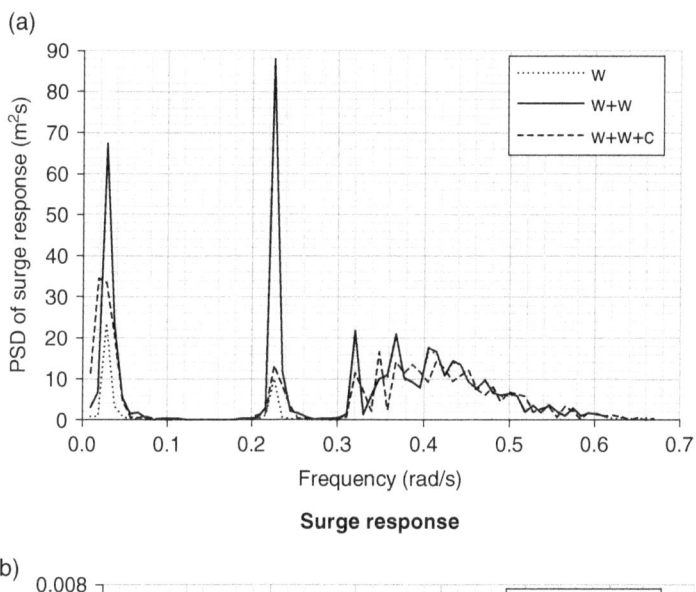

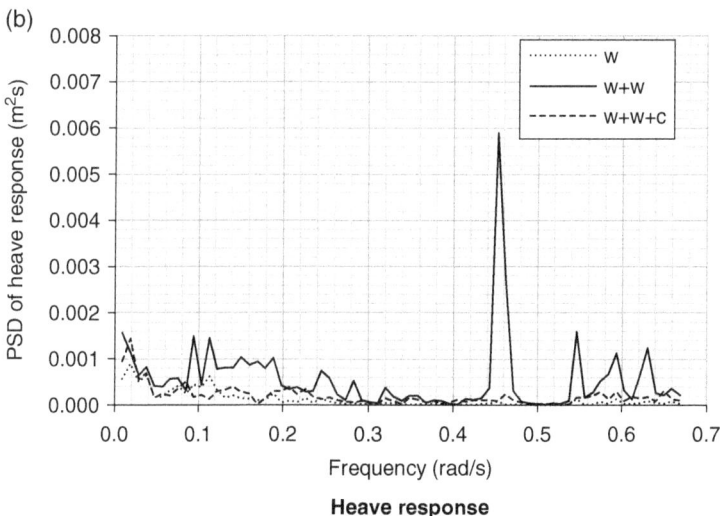

Figure 3.18 Deck response with high sea conditions (w – waves, w+w – waves+wind, w+w+c – waves+wind+current).

(c)

<chart: PSD of pitch response (deg²s) vs Frequency (rad/s), showing curves for W, W+W, and W+W+C>

Pitch response

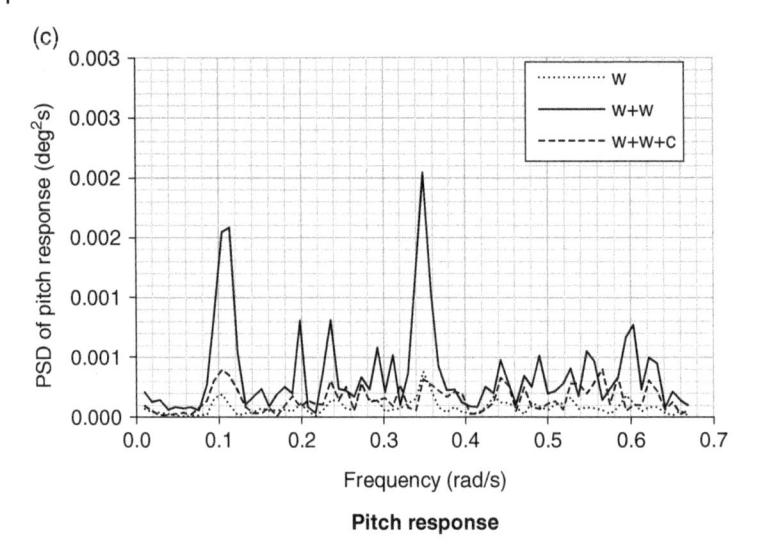

Figure 3.18 (Continued)

The addition of wind has a significant effect on increasing the pitch response. The PSD of the pitch response shows several peaks, with the maximum peak occurring close to one-fourth of the heave natural frequency. Due to the coupling effect, the current action reduces the total pitch response (Chandrasekaran and Nagavinothini 2019a, b).

An increased deck response is observed in high sea conditions as compared to moderate sea conditions. From moderate to high sea conditions, the surge response increases by 2.97 times, heave by 5.10 times, and pitch by 1.54 times, given the combined action of wind and waves. Despite the increase in the responses, the total heave response is only about 1.26% of the total surge response given the combined action of wind and waves. Similarly, given the combined action of wind, waves, and current, the surge, heave, and pitch deck responses increase by 2.62 times, 5.58 times, and 4.53 times, respectively, from moderate to high sea conditions. Though the total response increases with the roughness of the sea conditions in different DOF, the effect of wind on the structural response is reduced in very high sea conditions. Apart from this, the platform is found to be stable even with very high sea conditions under different loads, as seen from the surge phase plots in Figure 3.19. Similar responses also occur in other DOF. The stability of the platform is not affected even by very high sea conditions and the combined actions of wind, waves, and current. This response behavior is mainly attributed to the novel structural form of the triceratops, which makes it suitable for ultra-deepwater conditions.

(a)

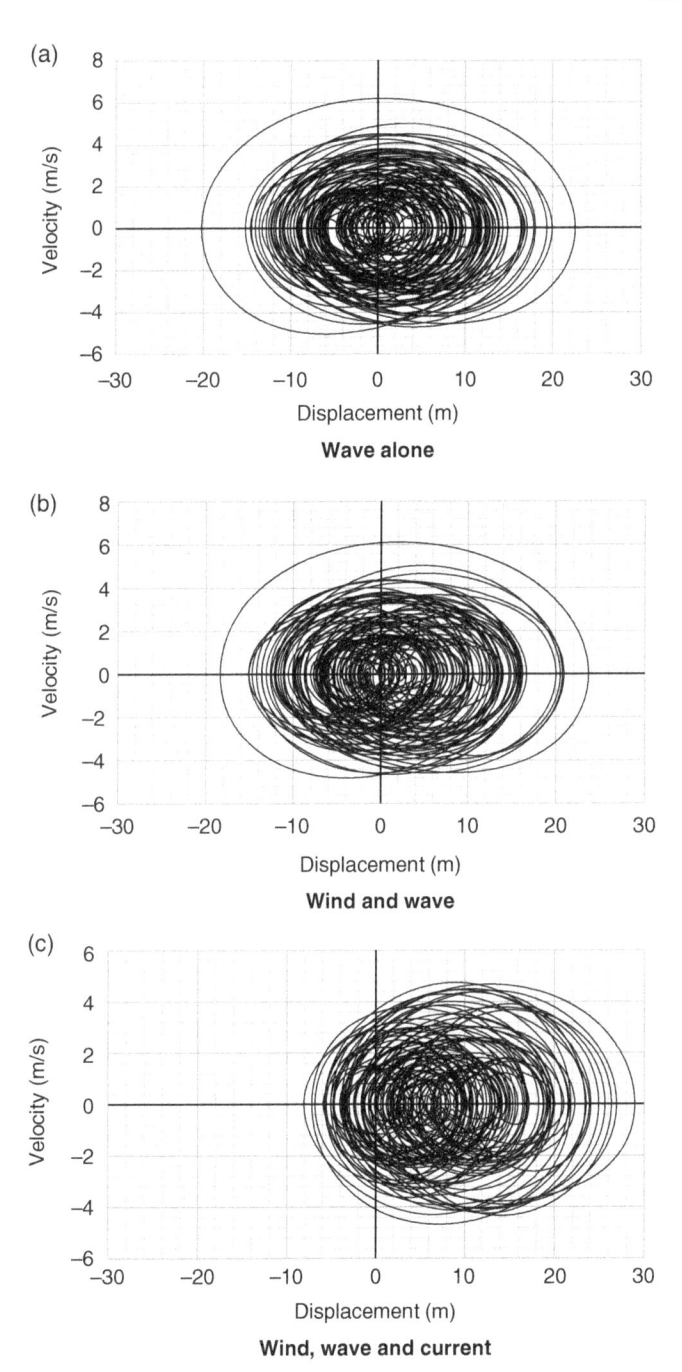

(b)

(c)

Figure 3.19 Phase plots in the surge DOF with very high sea conditions.

3.6.2 Buoyant Leg Response

From the behavior of the buoyant legs under different environmental loading conditions, the effect of the addition of wind loads and current loads can be understood clearly. The effect of wind and current loads is significant given high sea conditions. The addition of wind load increases the surge response in the buoyant legs by 62%. This occurs mainly due to the monolithic transfer of the surge response between the deck and the buoyant legs. However, the response is maximized in the case of the buoyant legs due to the coupling between the surge and pitch DOF. The pitch response developed in the buoyant legs due to the action of the wave load near MSL leads to an increase in the surge response of the buoyant legs, as the center of gravity of the buoyant legs is very close to the keel. The current action increases the surge response with an excessive shift in the mean value. The variation in the heave response of the three buoyant legs ranges from 30 to 40%, which is responsible for the pitch response in the deck even given the partial restraint produced by the ball joints. The wind action increases the heave response by 44%, but the current action decreases the heave response. The pitch response is increased by 37% with the addition of wind, due to the coupling between the surge and pitch DOF. As the current load acts on the buoyant legs up to a water depth of 200 m, this reduces the additional moment caused by the wind load on the buoyant legs up to 30%.

In the surge PSD plots of the buoyant legs given high sea conditions (Figure 3.20), the peak values are close to the surge natural frequency under different environmental loading conditions. The variation in spectral energy due to the addition of

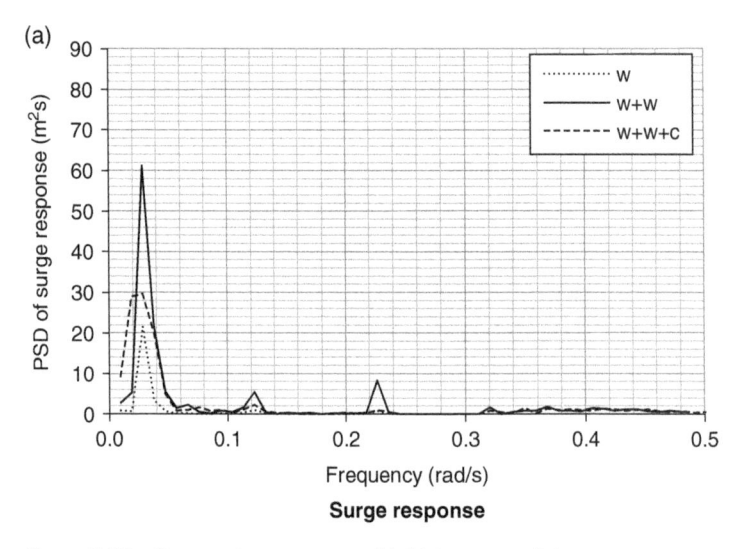

Figure 3.20 Buoyant leg response with high sea conditions.

(b)

Heave response

(c)

Pitch response

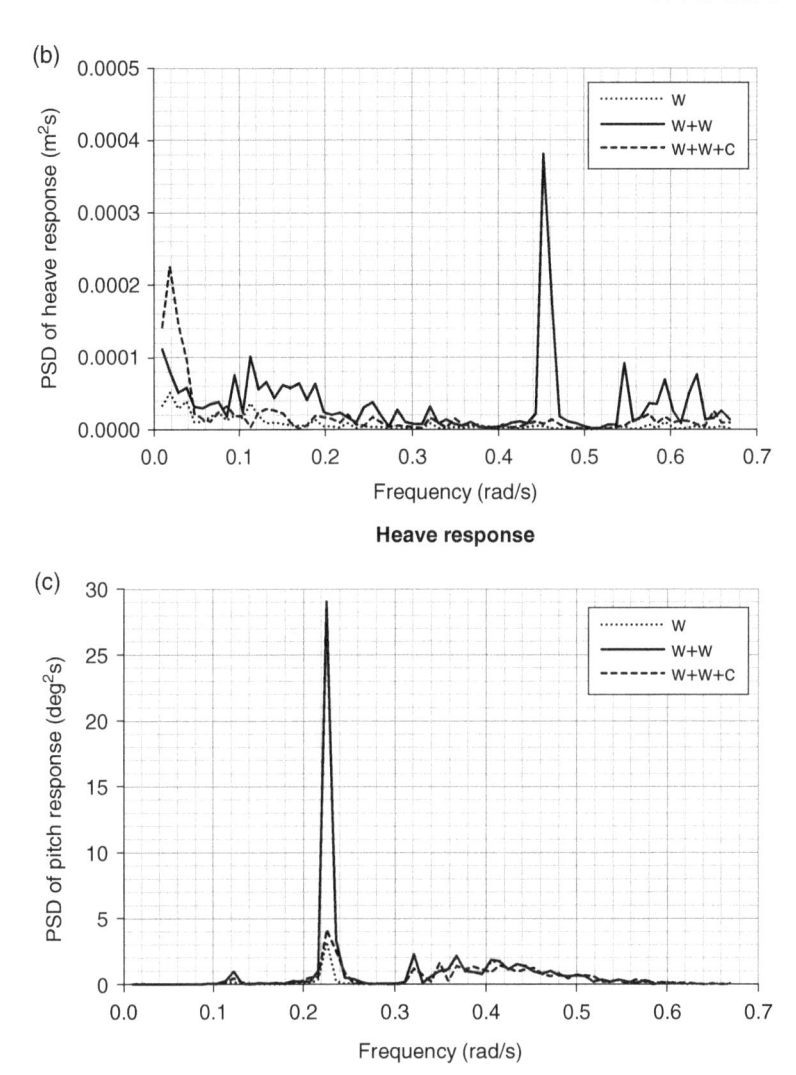

Figure 3.20 (Continued)

current and wind takes place very close to the surge natural frequency. In the heave response, the wind action increases the spectral energy close to half of the heave natural frequency, and the current action increases the response at lower frequencies. The wind action increases the spectral energy of the buoyant legs significantly at one-fourth of the pitch natural frequency. The current action reduces the spectral peak value due to the reduction in the pitch response.

3.6.3 Tether Tension Variation

The wind action increases the tether tension variation from 5.73 to 5.94% in moderate sea conditions, as given in Table 3.7. In the case of high and very high sea conditions, the tether tension variation is reduced. The dynamic tension variation increases with an increase in the roughness of the sea conditions, but the effect of wind is marginal. The maximum stress developed in the tethers even with very high sea conditions is well below the yield stress, which confirms that tether failure will not occur due to the development of high axial stress. Given the combined action of wind, waves, and current, high tether tension variation is observed during the initial excitation under different sea conditions. The addition of current increases the tether tension variation in moderate and high sea conditions. Unlike the tether response to the combined action of wind and waves, the tether tension variation with the combined action of wind, waves, and current decreases with increased roughness of the sea conditions. The wind load slightly increases the spectral energy of the tether tension. The peak value occurs close to the peak wave frequency. The current action enhances the tension spectrum by 21% at lower frequencies. A shift in the peaks occurs at higher frequencies due to the current action, as seen in Figure 3.21. The effect of wind and current in the service life of the tethers is very marginal. In the present case, the service life of the tethers varies from 13.83 years to 14.21 years with different loading and sea conditions (Chandrasekaran and Nagavinothini 2018a, b).

Table 3.7 Tether tension variation with combined actions of wind, waves, and current.

Sea condition	Statistics (MN)	Waves alone	Waves + Wind	Wind + Waves + Current
Moderate	Maximum	29.474	28.882	30.765
	Minimum	26.301	25.602	23.773
	Mean	27.671	27.624	27.678
	Tension variation (%)	5.73	5.94	12.63
High	Maximum	29.958	28.936	30.173
	Minimum	25.709	25.494	24.580
	Mean	27.670	27.583	27.640
	Tension variation (%)	7.68	6.24	10.12
Very high	Maximum	30.388	30.012	29.958
	Minimum	24.849	24.633	24.741
	Mean	27.751	27.637	27.674
	Tension variation (%)	9.98	9.73	9.43

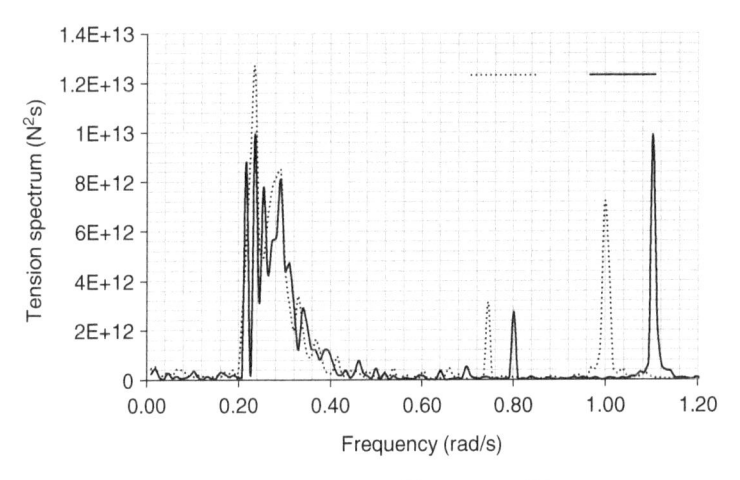

Figure 3.21 Tension spectrum with very high sea conditions.

3.7 Summary

Triceratops platforms are advantageous in the case of deepwater and ultra-deepwater applications. The heave response of a triceratops is only about 1.0% of the surge response, which shows the adequacy of the platform for drilling and production operations. The presence of ball joints is an additional advantage in the triceratops: they restrict the transfer of rotational motion from the buoyant legs to the deck in response to wave actions at different wave approach angles. The response of the triceratops in the active DOF increases with increased roughness of the sea conditions. The maximum spectral energy in different DOF occurs close to multiples or fractions of the natural frequency of the platform due to the complex behavior of the platform in response to the wave action. The service life of the tethers is reduced as a result of the action of random waves. The addition of wind load and current load increases the offset of the platform, and the platform starts oscillating in the new position. The surge PSD plots for the combined actions of wind and waves show peaks at the surge natural frequency and one-fourth of the pitch natural frequency, confirming the coupling between the surge and pitch DOF. The maximum spectral energy given combined wind, wave, and current actions occurs at lower frequencies. The coupling between the surge and heave DOF also increases the heave response to the action of wind and current. The service life of the tethers varies marginally based on the action of wind and current loads.

4

Triceratops Under Special Loads

Summary

This chapter describes the response of a triceratops platform to continuous ice crushing for an uncoupled dynamic analysis of offshore structures. It also includes the detailed response of an offshore triceratops to continuous ice crushing and a tether analysis given different sea ice conditions. This chapter also presents a response analysis for a triceratops under an impact load arising due to ship–platform collisions. The behavior of the deck plates in the presence of a hydrocarbon fire is also presented.

4.1 Introduction

In addition to environmental loads, offshore structures are also subjected to several other special loads, which sometimes may endanger the strength and stability of the whole structure. The challenges also increase with increased water depth. Some of the special loads that act on offshore structures include the following:

- Environmental loads arising due to ice, earthquakes, tides, and marine growth
- Loads due to temperature variations and seafloor movement
- Accidental loads arising due to ship–platform collisions, dropped objects, fires, explosions, changes of intended pressure differences during drilling, and failure of mooring lines on compliant structures

This chapter mainly deals with a response analysis of an offshore triceratops under ice load due to continuous ice crushing, impact loads arising due to ship–platform collisions, and fire loads on the deck of the platform. The numerical analysis methodology suggested herein can be applied to investigate the dynamic response of any compliant platforms.

Offshore Compliant Platforms: Analysis, Design, and Experimental Studies,
First Edition. Srinivasan Chandrasekaran and R. Nagavinothini.
© 2020 John Wiley & Sons Ltd.
This Work is a co-publication between John Wiley & Sons Ltd and ASME Press.

4.1.1 Ice Load

Sea ice develops during winter due to cooling of the surface seawater, which covers 15% of the world's oceans. The ice thickness and coverage increase progressively during the winter season, and the hardness of the ice increases with its age. During the spring season, the first-year ice, which is up to 3.0 m thick, melts due to increased temperature; but the multiyear ice in the northern and southern ocean regions remains unaffected throughout the year (Mahoney et al. 2004). During ice formation, needle-like crystals of ice develop first, which are then modified into an elastic ice crust by further freezing. Exposure of the ice to wind, waves, and currents leads to deformation of the ice along with an increase in the brittleness of the ice crust. It also results in the separation of sea ice into circular pieces of diameter up to 20 m. Circular ice pieces of diameter up to 3.0 m are called *pancake ice*, and the larger pieces are called *ice cakes*. Pancake ice causes impact forces on offshore structures, which increase with increased wave height and current fields (Sun and Shen 2012). Based on the environmental conditions, the ice cakes freeze together, resulting in the formation of a large, continuous layer of sheet ice sheet called an *ice floe*. Ice floes freezing together results in the formation of *ice fields* covering more than 10 km. Under the action of sea ice floes, structural motion is highly affected by the wavelength (McGovern and Bai 2014). Also, the drift velocity controls the ice floe's impact on offshore structures. The continuous action of wind, waves, and current transforms this continuous flat ice sheet into pressure ice fields with rough surfaces. Also, the piling up of ice sheets in an irregular manner results in the formation of ice pressure ridges. The ice sheet that remains attached to the shore developed during winter is called *shore-fast ice*. It detaches from the shore during the spring season and is then referred to as *pack ice*. The pack ice has a width extending up to 10 km, and it is in continuous motion with velocities similar to the current velocity.

Apart from seawater ice, ice shelves that originate from the land by compaction of snow layers and freezing of fresh water are called *glaciers*. *Icebergs* then form due to the flow of glaciers followed by chunks of ice breaking due to the buoyancy of water. The direction and amplitude of wind and currents govern the velocity of the icebergs in a particular location. The temperature variations above and below the water surface cause non-uniform melting of icebergs, which results in the icebergs tilting, capsizing, and breaking. Breakage of icebergs leads to the formation of smaller bergs called *growlers* or *bergy bits*. The design of offshore structures is mainly governed by ice sheets, pack ice, and icebergs (Reddy and Swamidas 2016).

The considerable increase in oil drilling and production in the Arctic region in the recent past and the hindrance to normal operations caused by moving ice sheets demands an innovative structural form for offshore applications. Also, cold environmental conditions pose additional challenges by affecting the material properties of the structure. Offshore platforms to be installed in ice-infested cold regions should be analyzed for dynamic effects caused by the random and cyclic characteristics of the ice loads (Shih 1991). The ice load acting on an offshore structure depends upon the following factors:

- Structural geometry of the platform
- Location and environmental conditions
- Ice properties such as thickness, velocity, and crushing strength
- Ice-structure interaction phenomenon

When an ice sheet hits a vertical structure due to the action of wind, waves, and currents, continuous failure of the ice occurs, which results in a horizontal force on the structure. Under certain conditions, the ice-structure interaction may also result in transient vibrations due to pressure gradients developed from the continuous failure of the ice.

The most common failure cases that occur during ice-structure interactions are limit stress failure and limit force failure. In the case of a drifting ice sheet, ice sheet failure occurs at the ice-structure interface when the environmental forces acting on the ice are greater than the failure strength of the ice. The common modes of ice failure under limit stress failure conditions are buckling and crushing. In the case of limit force failure, the ice failure occurs far from the ice-structure interface, and the environmental forces acting on the ice sheet lead to ice ridge formation. This results in the movement of ice ridges around the structure and thus reduces the total ice force acting on the structure. The failure of an ice sheet depends upon the ice thickness, dimensions of the interacting structure, and strain rate. The different ice failure modes are as follows:

- Crushing
- Buckling
- Shear
- Radial and circumferential cracking
- Creep
- Spalling

When the ratio of ice thickness to leg diameter is lower, ice failure occurs due to creep and crushing modes at lower and higher strain rates, respectively. In the case of a higher ratio of ice thickness to leg diameter, the failure mode shifts to radial and circumferential cracking. Hence, suitable empirical relations should be used while calculating the ice forces on structures.

Level ice action induces random vibrations in offshore structures, and crushing ice failure causes maximum ice force. The ice-structure interaction itself is a complex phenomenon, which adds further complications to the analysis of structures in response to the ice action (Yue et al. 2001). Ice properties also affect the ice-induced vibration of structures (Karr et al. 1993). Increased ice velocity decreases the average crushing force and also increases the predominant frequency of the structural response up to a certain limit below the natural frequency of the structure. At low ice velocity, sawtooth-like oscillations occur in fixed structures. At high ice velocity, harmonic oscillations occur in flexible structures. Intermittent ice breakage and non-continuous contact between the ice sheet and the structure contribute to the highly complicated response of the structure under ice loads. Given crushing ice failure, three distinct ice force modes occur in response to varying ice speeds: quasi-static, steady condition, and random vibrations. Each ice force mode takes place under ductile, ductile-brittle transition, and brittle failure conditions (Yue et al. 2002). The random ice force and the corresponding vibration of the structure are shown in Figure 4.1. Thus, the global ice force is a function of the ice parameters and the compliance of the structure (Karna et al. 2006a, b).

A dynamic ice force model for crushing ice failure is developed by Karna et al. for both narrow and wide structures, based on the responses collected during ice interaction on the Norstromgrund lighthouse in Bohai Bay. It can be used as an effective tool to assess the behavior of structures in random ice force mode. As nonlinear effects were not considered in the development of the model, it is not

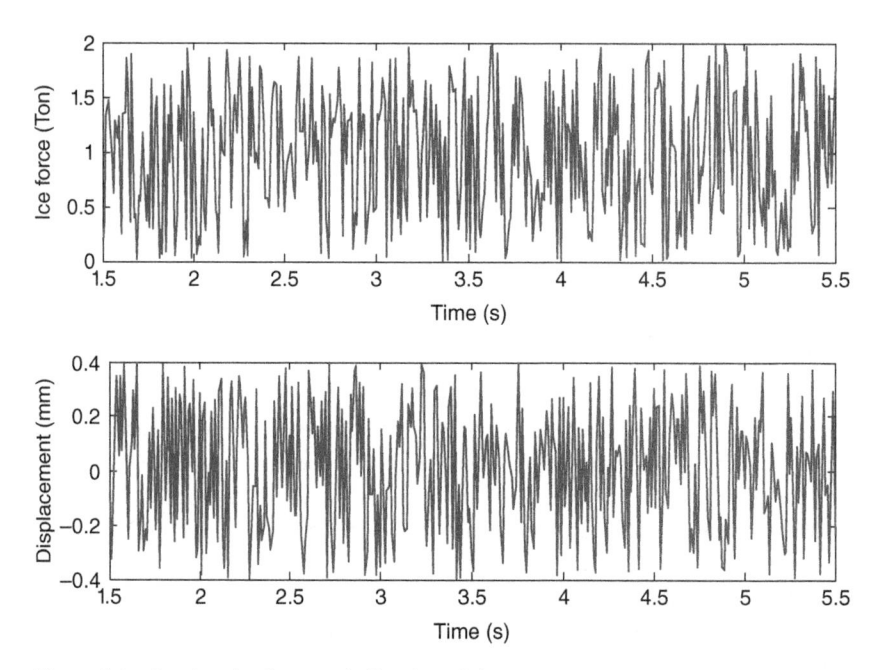

Figure 4.1 Random ice force and vibration of the structure.

applicable for intermittent ice-crushing phenomena (Karna et al. 2007). When the ice force frequency becomes equal to the natural frequency of the structure, the dynamic amplification of the structural force will be high; and greater dynamic force will occur when the ice force frequency becomes equal to the integer fraction of the natural frequency (Ziemer and Evers 2016). The study of ice–structure interactions becomes mandatory for the cost-effective structural design of offshore structures to be installed in ice-infested regions (Heinonen and Rissanen 2017).

4.1.2 Impact Load Due to Ship Platform Collisions

Offshore structures require servicing from large supply vessels due to their distance from the shore. It is very important to assess the collision resistance of offshore structures for a sensible design. Compliant platforms may be more at risk from impacts because the structures have very little or no redundancy. Also, the post-collapse strength of compliant platforms with hull structures is very low. Thus, a relatively small dent may be sufficient to eliminate the entire design safety factor of the structural member (Harding et al. 1983). In the case of a triceratops, the buoyant legs are prone to impact loads. Local and global deformations that develop due to impact loads on the buoyant legs may endanger the strength and stability of the entire platform. In the recent past, the risk associated with ship–platform collisions has grown substantially due to the increased number of oil and gas production platforms all over the world. Several collisions did not result in injuries or loss of life, but the economic losses were significant.

In order to reduce the failure of offshore platforms under accidental loads, standard regulations should be followed in the design of the structures (Amdahl and Eberg 1993). According to the NORSOK N-003 guidelines for production platforms, 5000-ton supply ships with speeds of not less than 2.0 m/s should be considered for impact analysis and design checks. The guidelines allow significant damage to the platform, but the design should be proper to avoid the progressive collapse of the platform. The same thing has been suggested by the regulations of the Norwegian Maritime Directorate and the Der Norske Veritas (DNV) standards. The design guidelines also suggest 4.0 MJ as the minimum collision energy for the design of offshore structures that may experience accidental collisions. From a statistical overview of collisions in the offshore industry, an increased number of collision events are happening between offshore structures and visiting vessels.

Further, there has been an increase in the size and weight of visiting vessels in the past 30 years. Vessels with greater weight are more capable of causing severe damage to structures. Recent collision events showed higher collision energy, but with a lesser probability of the design collision event, as indicated by the standard regulations.

130 | *Offshore Compliant Platforms*

The impact behavior of any structure will be highly affected by its material properties. In the case of numerical analysis of offshore structures under impact loads, it is necessary to define the material properties appropriately using a true stress–strain curve, which represents the condition of the material more accurately. The true stress–strain values can be calculated from the engineering stress–strain values using the following equations:

$$\sigma_t = \sigma_{eng}\left(1+\varepsilon_{eng}\right) \tag{4.1}$$

$$\varepsilon_t = \ln\left(1+\varepsilon_{eng}\right) \tag{4.2}$$

where σ_t is the true stress, σ_{eng} is the engineering stress, ε_t is the true strain, and ε_{eng} is the engineering strain. For example, the true stress–strain curve of AH36 marine steel is shown in Figure 4.2. The mechanical properties of AH36 marine steel are listed in Table 4.1.

In the impact analysis of offshore structures, it is vital to study local as well as global deformations in order to understand the exact behavior of the structures. During the design of the offshore platform, the deformation and behavior of the platform are of more concern than those of a ship. Hence, indenters with different shapes that resemble a ship can be modeled in the initial analysis stage. The stem of the ship can be modeled as a rectangular indenter. The shape can also be modified to a pointed end from the flat end. The different shapes of indenters that can be considered in the impact analysis are shown in Figure 4.3. Regarding the

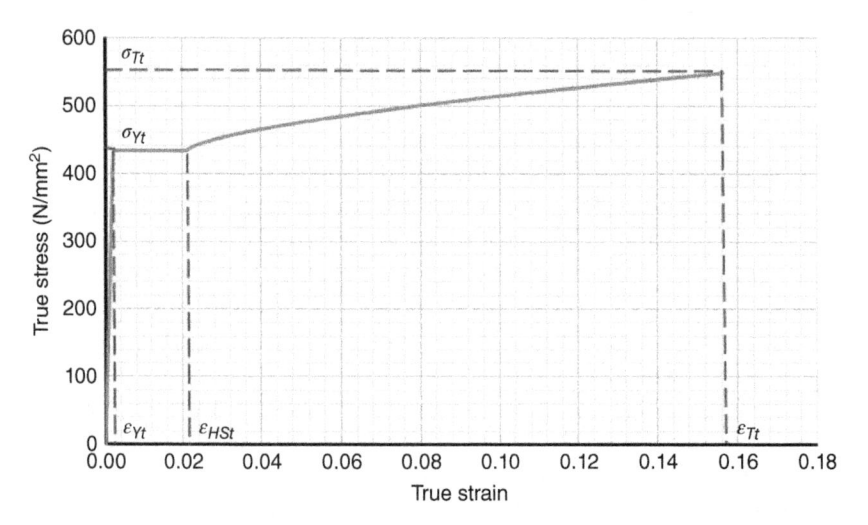

Figure 4.2 True stress–strain curve of AH36 grade steel.

Table 4.1 Mechanical properties of marine DH36 steel (Cho et al. 2015).

Mechanical properties	Value	Units
Yield strength	433	N/mm^2
Young's modulus	206 000	N/mm^2
Ultimate tensile strength	547	N/mm^2
Ultimate tensile strain	0.156	No unit
Hardening start strain	0.0214	No unit

Figure 4.3 Different shapes of indenters. (a) Rectangular (b) Knife-edge (c) Hemispherical

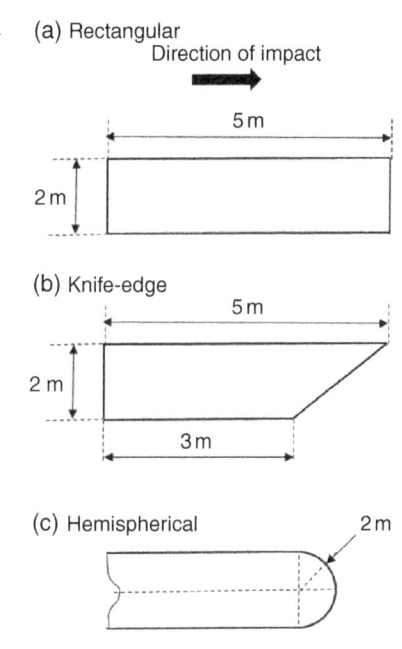

impact loads, robustness in the design of the exposed components should be incorporated by selecting appropriate materials with sufficient toughness, avoiding weak elements and critical components in vulnerable locations.

4.1.3 Hydrocarbon Fires

Hydrocarbon fires are among the most severe accidental loads on offshore platforms. Due to the degradation of material properties, fire accidents on offshore structures may result in the complete collapse of the structure. For the fire-resistant design of offshore structures, the nature of the fire load, performance requirements, and the response of the structure to a hydrocarbon fire become vital input. A few major accidents that have occurred in the past – Piper Alpha (1988), Petrobras (2001), Mumbai High North (2005), Ekofisk (1989), and Deepwater Horizon (2010) – reinforce the importance of safety and structural integrity for offshore platforms with regard to hydrocarbon fires (Manco et al. 2013; Rivera et al. 2014). A detailed study of the response of the structural components of an offshore platform is necessary, as it is very difficult to control a hydrocarbon fire after it breaks out. One of the major structural components in direct contact with the hydrocarbon fire is the deck. The deck is made from stiffened panels, which should be critically analyzed in the presence of a hydrocarbon fire. Normally, the behavior of any structure, either land-based or offshore, can be assessed by predicting the survival time of the structure after fire breaks out and before the occurrence of any form of structural failure using a standard time–temperature curve. The time–temperature curve for different fire conditions is shown in Figure 4.4.

Offshore structures, especially the topsides, are constructed with different types of steel. With respect to the grade of steel, the stress–strain characteristics vary significantly at elevated temperatures. An increase in temperature leads to thermal strains in the material, even in the absence of mechanical loading. So, the structural elements experience thermal strain without an increase in internal stresses due to higher temperatures. With increased temperatures, Young's modulus, stiffness, and the yield strength of structural steel decrease with or without the development of mechanical strains. On the other hand, the material ductility increases, showing an indication of strength development. In the case of mild carbon steel, the effective yield strength is reduced at higher temperatures (above 400 °C) at 2% strain, whereas the proportional limit and modulus of elasticity

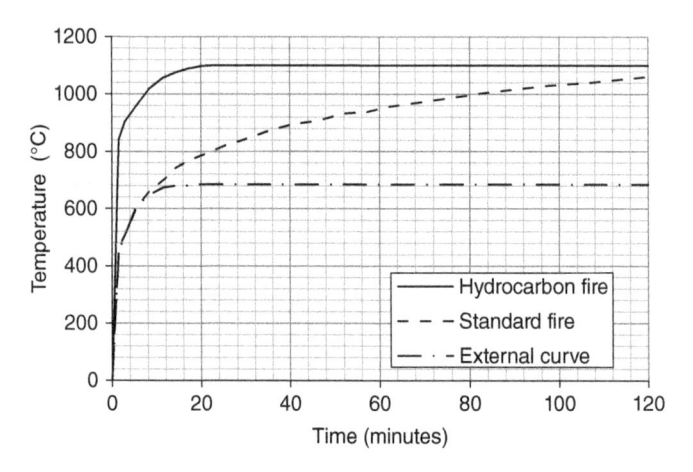

Figure 4.4 Time–temperature curves for different fire conditions.

decrease with temperatures over 100 °C, as shown in Figure 4.5 (Eurocode 3, Part 1–2). The principal mechanism that causes a reduction in the strength and stability of the structure during a fire is the release of potential energy. Through energy absorption, the denser internal structure of steel leads to a phase transformation at 730 °C. The variation of thermal conductivity, specific heat, thermal strain, and thermal expansion with increased temperatures are shown in Figure 4.6–4.8, respectively.

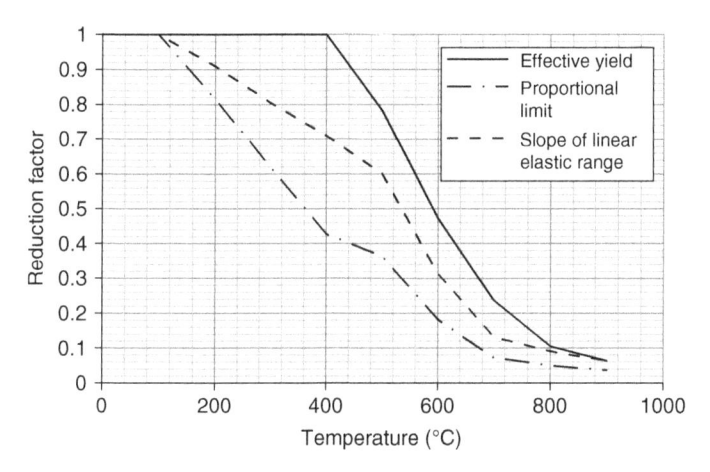

Figure 4.5 Reduction factors for yield strength, proportional limits, and linear elastic range for carbon steel.

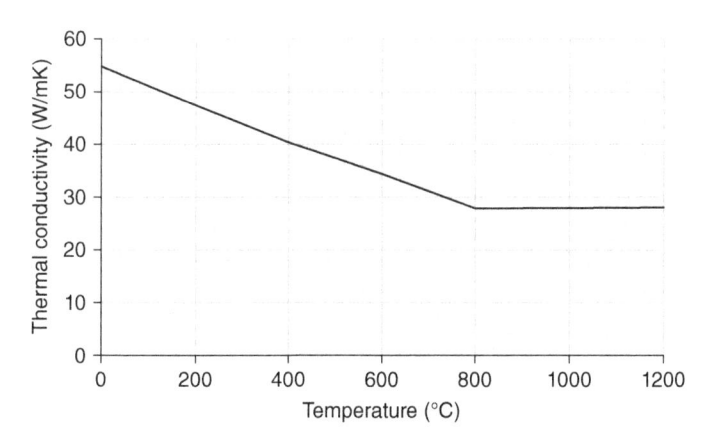

Figure 4.6 Variations in the thermal conductivity of carbon steel.

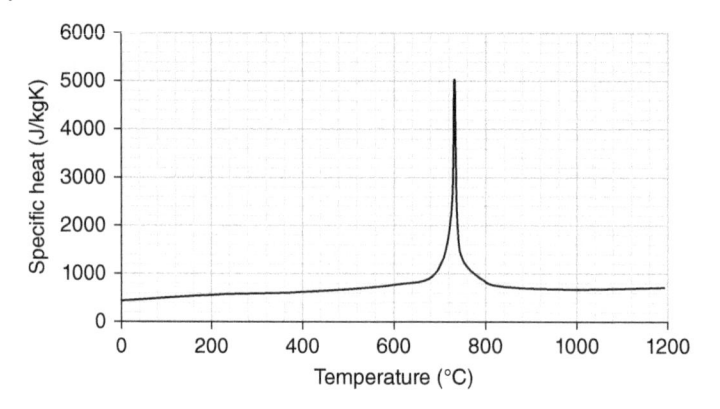

Figure 4.7 Variations in the specific heat of carbon steel.

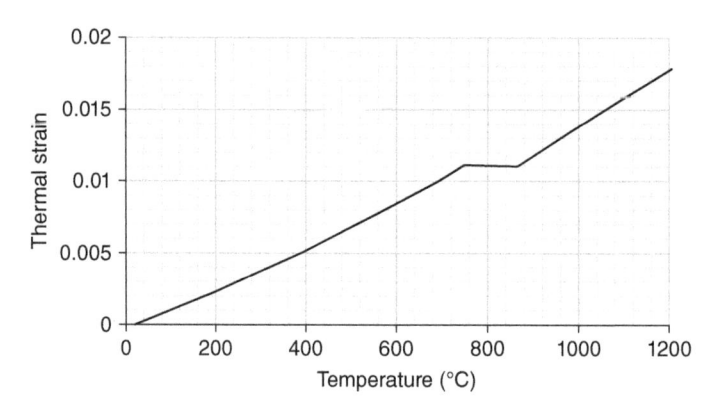

Figure 4.8 Variations in the thermal strain of carbon steel.

4.2 Continuous Ice Crushing

The major cause of ice drifting is the action of wind, waves, and currents on ice sheets. Also, ice velocity depends upon its location. Given the availability of sufficient energy to cause ice failure, limiting the stress induced by that failure governs the necessity of including ice loads in the design of offshore structures. Under these conditions, *ice force* can be defined as the force required to cause ice failure at the ice–structure contact region. The major factor that limits the maximum ice force acting on any structure is the ice failure mechanism. The ice failure mechanism, in turn, depends upon ice parameters such as ice thickness, ice velocity, width of the ice plate, and shape of the structure. When an ice sheet interacts with a compliant structure, ice failure occurs due to ductile and brittle modes given low and high velocities, respectively. As a result, the continuous ice crushing phenomenon occurs, given high ice velocity. Ice crushing is one of the common ice failure mechanisms in ice sheets, which results in maximum ice force on structures. It occurs when a sheet of ice hits a vertical-sided structure with moderate to high ice velocity. During this process, horizontal cracks form on the ice sheets at the contact zone, leading to pulverization of the ice sheet. The crushed ice particles in the vicinity of the structure pile up and slide around the structure, resulting in the structure vibrating. The ice forces acting on a structure under crushing ice failure are a function of the ice strength, which depends upon the ice thickness and formation. Continuous ice crushing during ice–structure interaction results in non-uniform partial contact, and non-simultaneous pressure on the contact area. The ice force–time history will have waveforms with randomly distributed amplitudes and periods. Thus, ice force can be designated as a stochastic process and described using a frequency spectrum. The uncoupled time-dependent load can be used in the dynamic analysis of structures because the transition between the different modes of failure is not completely established.

4.2.1 The Korzhavin Equation

Sea ice exists very close to its melting point. Thus, the mechanical properties of the ice are highly affected by temperature. Several empirical equations have been developed for the calculation of the mechanical properties of ice. The maximum crushing ice force is calculated by multiplying the effective ice pressure and the contact area. The effective ice crushing pressure depends upon several factors such as aspect ratio, confinement within the ice sheet, scale, and degree of contact between the ice and the structure (Sodhi and Haehnel 2003). One of the initial approaches for calculating the limit of the ice load acting on a vertical structure is the Korzhavin equation (McCoy et al. 2014). Though it is one of the old approaches, it is still followed with amendments made with modification factors. The equation is given by:

$$F = a_1 a_2 a_3 h w \sigma_c \tag{4.3}$$

where a_1 is the shape factor (0.9 for circular members), a_2 is the contact factor (0.5 for moving ice), a_3 is the aspect ratio factor, σ_c is the crushing strength of the ice in MPa, h is the thickness of the ice in meters, and w is the projected width of the structure in meters. The important parameter that affects the crushing strength of the ice is the temperature. Under spring conditions in the Beaufort Sea, when the temperature is close to the melting point of the ice, the recorded average crushing strength of ice is 1.5 MPa. During the coldest time of the year, the crushing strength of ice is 3.0 MPa. These can be considered the normal and extreme ice sea conditions, respectively.

4.2.2 Continuous Ice Crushing Spectrum

Continuous ice crushing occurs due to high ice speed, and ice crushing strength does not influence the response of the structure. Under such conditions, the feedback mechanism developed by the compliant structure in response to the ice load action becomes negligible. This makes the spectral model developed from the response data of existing structures more useful for the analysis of offshore compliant structures. Karna et al. developed a continuous crushing ice force spectrum for ice speed ranging from 0.04 to 0.35 m/s (Karna et al. 2007). The nondimensional spectral density function is given by:

$$\bar{G}_n(f) = \frac{af}{1 + k_s a^{1.5} f^2} \quad (4.4)$$

where $a = bv^{-0.6}$, v is the ice velocity in m/s, b and k_s are the experimental parameters, and f is the frequency in hertz. Now the spectral density function is given by

$$G_n(f) = \frac{\sigma_n^2 \bar{G}_n(f)}{f} \quad (4.5)$$

where σ_n^2 is the variance of the local force. The mean ice force and standard deviation are calculated from the following set of equations:

$$\sigma_n = \frac{I_n}{1 + kI_n} F_n^{max} \quad (4.6)$$

$$F_n^{mean} = \frac{F_n^{max}}{1 + kI_n} \quad (4.7)$$

where I_n is an intensity parameter that varies from 0.2 to 0.5, k is the probability of exceeding the event under consideration, F^{mean} is the mean ice force, and F^{max} is the maximum ice force. Given 1.5 m ice thickness and 1.5 MPa ice crushing strength, the ice force spectrum with different ice velocities is shown in Figure 4.9. The peak occurs at very low frequencies, and the variation in the spectral energy is concentrated only at frequencies below 0.3 Hz with the variation in ice velocity. This shows that the effect of ice velocity on the time-varying component of the ice force is much less, and thus the major factor that dominates the response of the structure is the mean ice force.

Spectral plots of different maximum ice forces for 0.2 m/s velocity are shown in Figure 4.10. The significant variation in the ice force spectrum shows that the response of the structure to ice action will be highly affected by the maximum ice force. From the spectra developed for different ice properties, the ice load time

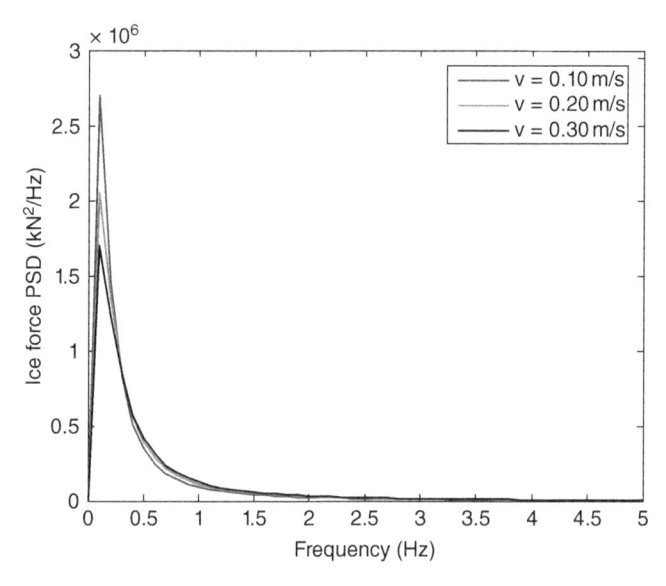

Figure 4.9 Spectral density plot given different ice velocities.

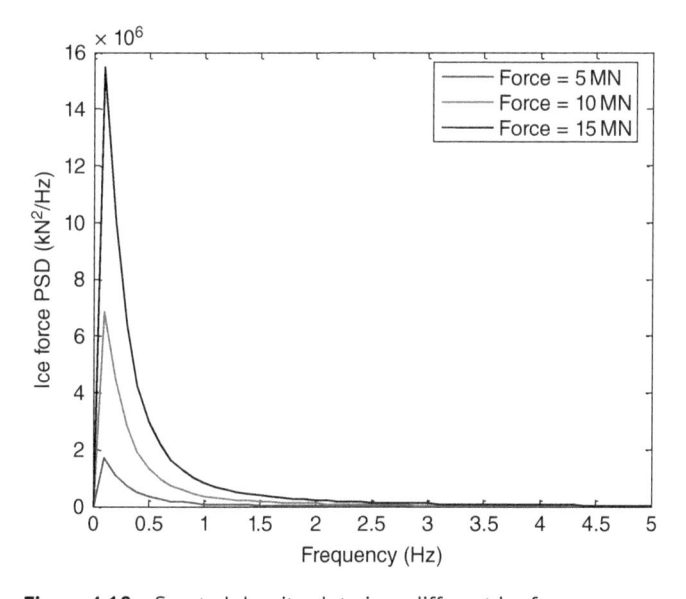

Figure 4.10 Spectral density plot given different ice forces.

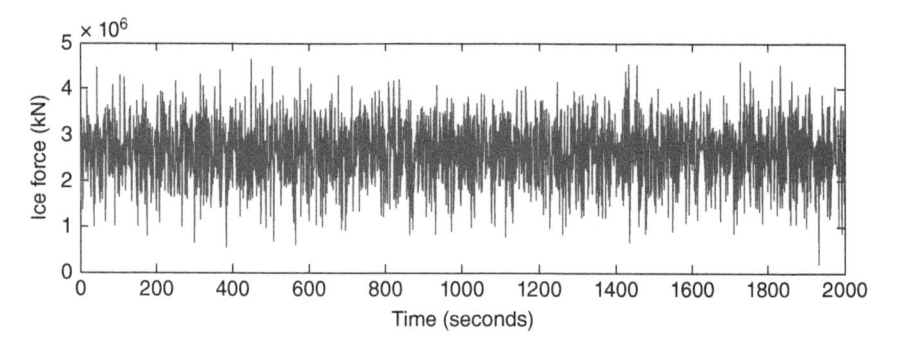

Figure 4.11 Ice force-time history.

history can be developed by using inverse fast Fourier transform (IFFT). The developed ice load-time history can be used to carry out the numerical analysis of offshore structures in the time domain. An example of an ice load-time history for 0.2 m/s ice velocity, 1.5 MPa ice crushing strength, and 0.5 m ice thickness is shown in Figure 4.11.

4.3 Response to Continuous Ice Crushing

The response of any offshore structure should be carefully investigated for different load cases in order to understand its exact behavior given real sea conditions. In the case of a response analysis of an offshore structure given an ice load action, the two load cases mentioned in Table 4.2 can be considered. The maximum ice load for the different load cases is calculated using the Korzhavin equation for the ice load action on a buoyant leg of the triceratops discussed in Chapter 3. In order to examine the behavior of the triceratops in response to continuous ice crushing, the ice spectrum model is developed for both cases, from which the ice force-time history is obtained. The ice force is applied as an external force acting on one or two buoyant legs for the numerical analysis. In the case of a three-legged structure like a triceratops, the maximum ice force occurs when ice acts

Table 4.2 Ice sea conditions.

Ice sea condition	Ice thickness (m)	Crushing strength (MPa)	Ice velocity (m/s)	Maximum ice load (kN)
Normal	0.5	1.5	0.2	5468.10
Extreme	1	3	0.2	23 382.70

on two buoyant legs simultaneously, and the total load is twice the maximum load on one leg.

4.3.1 Response to Ice Loads

4.3.1.1 Deck and Buoyant Leg Responses

When an ice load acts on two buoyant legs, the complex behavior of the triceratops shows significant responses in both the deck and buoyant legs in all degrees of freedom (DOF). The ball joints transfer the translational motion and restrict the transfer of rotational motion between the deck and the buoyant legs. Though the ice load on two buoyant legs acts in the positive x direction, it induces transverse vibration in the buoyant legs, which is transferred to the deck. This causes a significant response in all DOF. The response statistics under normal and extreme ice sea conditions are shown in Table 4.3.

The response of the triceratops increases with increased maximum ice load under different load cases. The ice load action causes a shift in the mean position in all DOF. The difference in the heave response in the three buoyant legs occurs due to the ice action, and this induces additional moment in the deck, resulting in roll and pitch responses. However, the heave response in a triceratops is very low even under extreme ice load action, thus confirming the operational advantage of the platform. There is an increase in the response of the deck in all DOF with increased ice forces, which is evident from the increased standard deviation values. The maximum surge response in extreme sea conditions is only about 9.66% of the draft of the structure.

Table 4.3 Deck response to different sea conditions.

Sea condition	Statistics	Surge (m)	Sway (m)	Heave (m)	Roll (deg)	Pitch (deg)	Yaw (deg)
Normal	Maximum	4.540	0.475	0.000	0.456	0.047	6.469
	Minimum	0.000	−0.594	−0.180	−0.003	−0.241	0.000
	Mean	2.489	−0.055	−0.062	0.151	−0.077	3.623
	SD	0.902	0.281	0.036	0.096	0.053	0.981
	RMS	2.648	0.286	0.072	0.179	0.093	3.754
Extreme	Maximum	14.879	1.270	0.000	6.958	0.042	23.997
	Minimum	0.000	−3.093	−2.696	0.000	−3.927	0.000
	Mean	10.273	−1.023	−0.967	2.454	−1.344	16.702
	SD	1.913	0.825	0.387	1.007	0.575	2.835
	RMS	10.450	1.314	1.041	2.653	1.462	16.941

The power spectral density (PSD) plots of the deck and buoyant leg responses in all DOF are shown in Figure 4.12. In the surge DOF, the first peak occurs at a frequency of 0.028 rad/s, which is equal to the natural surge frequency. The first peak is dominant in buoyant leg 1. The second peak occurs at a frequency of 0.123 rad/s (close to four times the natural surge frequency), and it occurs only in the buoyant legs. The third peak occurs at a frequency of 0.235 rad/s, in the neighborhood of one-fourth of the natural pitch frequency. In the sway response, dominant peaks are observed at 0.028, 0.123, and 0.226 rad/s. The third peak is dominant in the deck, and the first two peaks occur at the buoyant legs. Unlike the surge and sway responses, the heave response shows only two peaks at the frequencies 0.123 and 0.226 rad/s. The maximum response in the roll DOF is observed only in the buoyant legs. The deck response in the pitch DOF is very low compared to that of the buoyant legs, which again shows the efficiency of the ball joints in restraining the rotational DOF from the buoyant legs to the deck. In the yaw response, only one peak is observed, at a frequency of 0.028 rad/s in the deck. It shows the dominance of the natural surge frequency in the response of the platform in a horizontal plane. Similarly, the dominant frequency in the vertical plane is 0.226 rad/s, close to one-fourth of the natural pitch frequency. The compliant DOF of the platform are highly affected by the action of ice forces. Other DOF are also activated due to the action of ice loads in two buoyant legs and the coupling between the surge and pitch DOF.

4.3.1.2 Tether Response

Under normal conditions, the maximum tether tension is observed in buoyant leg 1 and buoyant leg 3. In extreme sea conditions, the maximum tether tension variation is observed in buoyant leg 1: about 5.50%. The mean value shift from the initial tether tension in buoyant leg 1 is found to be 0.99 MN, and the mean shifts in the tethers present in the other buoyant legs are comparatively less. The PSD plots of tether tension in buoyant leg 1 are shown in Figure 4.13. Under normal sea conditions, the first peak occurs at a frequency of 0.123 rad/s, four times the natural surge frequency. The second peak occurs at one-fourth of the natural pitch frequency.

4.3.2 Effect of Ice Parameters

4.3.2.1 Ice Thickness

Ice thickness increases the limit of an ice load acting on the structure linearly, which in turn increases the response of the structure. The minimum ice thickness for crushing ice failure to happen is 0.20 m. The maximum ice thickness observed in the Arctic region in recent years is around 1.0 m. The responses of the deck and

(a) Surge response

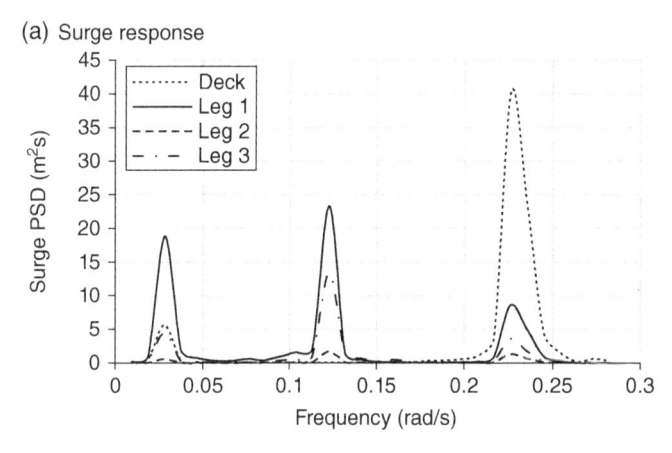

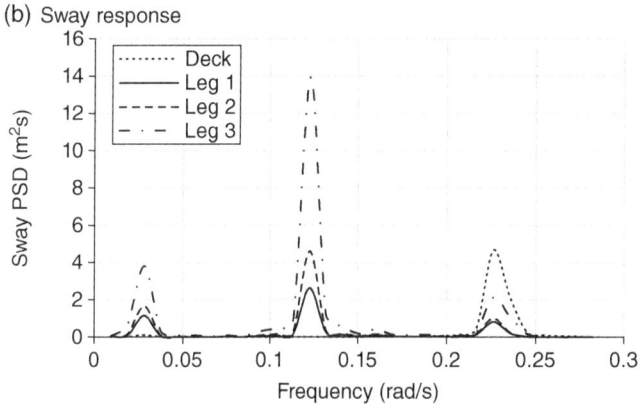

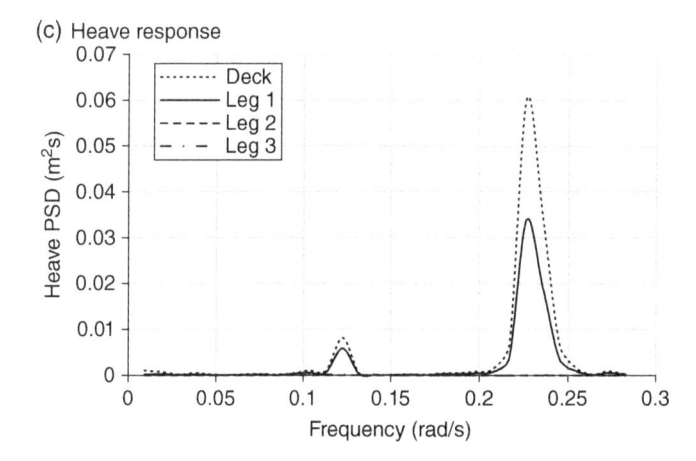

(b) Sway response

(c) Heave response

Figure 4.12 PSD plots for normal ice sea conditions with ice load on two buoyant legs.

142 *Offshore Compliant Platforms*

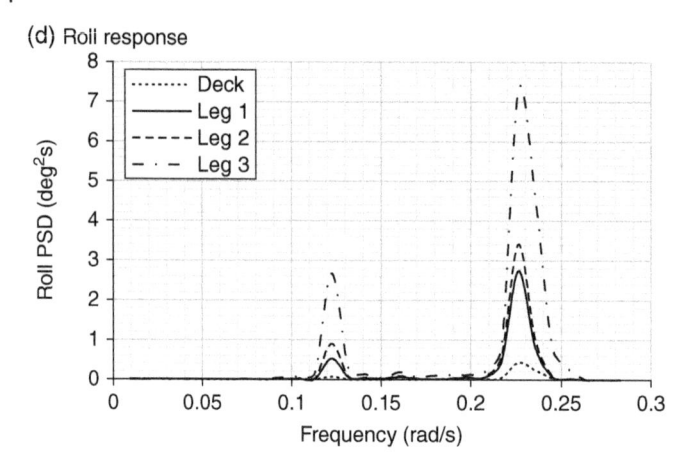

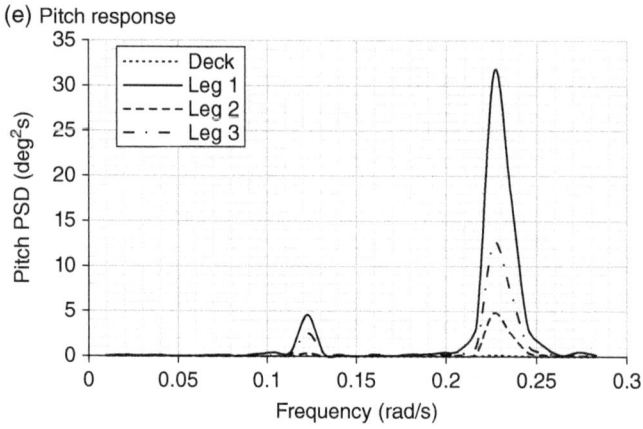

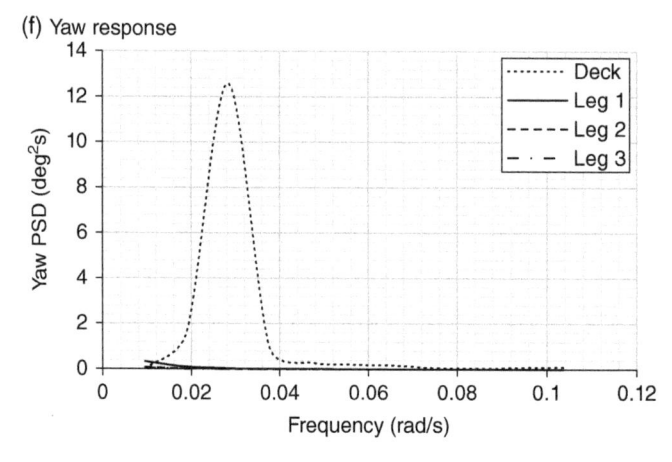

Figure 4.12 (Continued)

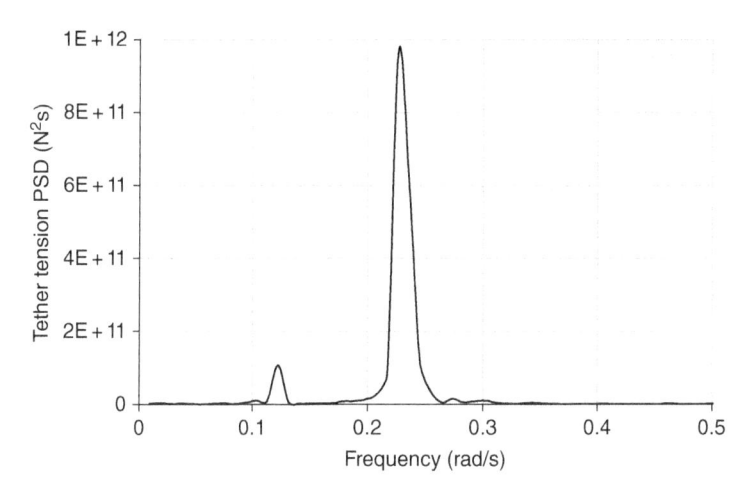

Figure 4.13 PSD plots of tether tension variation in normal sea conditions.

buoyant legs of the triceratops increase with increased ice thickness. The change in the total deck response of the triceratops due to different ice velocities is shown in Figure 4.14. With increased ice thickness, the spectral energy for different DOF also increases. The maximum tether tension increases with increased ice thickness. There is also an increase in the mean shift from the initial tether tension increase with increased ice thickness. At maximum ice thickness, the mean shift is about 0.28 MN.

4.3.2.2 Ice Crushing Strength

Similar to the ice thickness, the limit of the ice load increases linearly with increased ice crushing strength. The response in all DOF increases with increased ice crushing strength. The maximum surge response increases by about 14.14, 37.72, and 2.7% with the successive increase in ice crushing strength from 1.5–3.00 MPa. The shift in the mean value of the surge response also increases with increased ice strength. The total heave response at the maximum ice crushing strength is only about 7% of the total surge response. The mean values of the roll and pitch responses are found to increase in the positive and negative directions, respectively, in all cases. Increased total response of the deck with increased ice crushing strength is shown in Figure 4.15. The maximum tether tension increases with increased ice crushing strength up to 2.5 MPa and then decreases. A similar trend is observed in the heave, roll, and pitch responses of the buoyant legs. The maximum tether tension at 2.5 MPa crushing strength is about 28.5 MN. It occurs due to increased responses in all DOF at 2.5 MPa ice crushing strength.

(a) Surge, sway and yaw response

(b) Heave, roll and pitch response

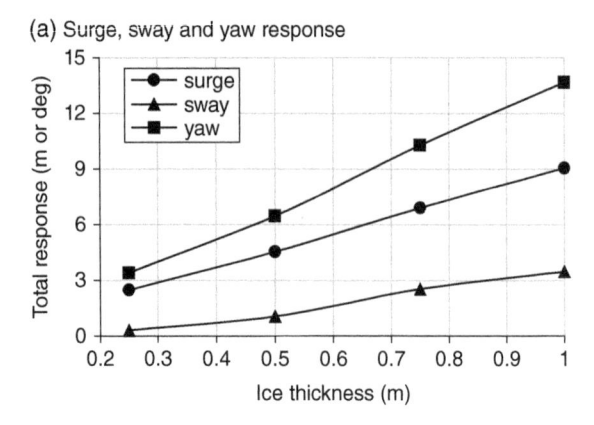

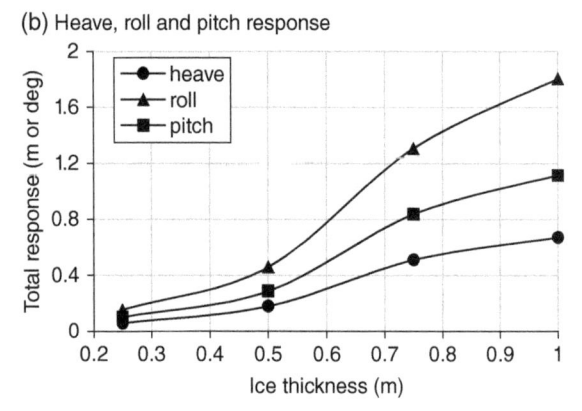

Figure 4.14 Total deck response for different ice thicknesses.

4.3.2.3 Ice Velocity

The ice force spectrum was developed for the velocity range of 0.04 m/d and 0.35 m/s. The total deck response with varying ice velocities given different DOF is shown in Figure 4.16. As changing ice velocity does not alter the time-varying component of the ice load–time history, the variation in the deck response in different DOF is much less. The ice forces acting in the positive x direction lead to maximum surge, heave, and pitch responses at high ice velocities. The sway, roll, and yaw responses are activated at lower ice velocities. Increased tether tension variation and tether pullout in two buoyant legs at the same time may endanger the stability of the platform. The maximum tether tension variation is 1.60% given very high ice velocity.

(a) Surge, sway and yaw response

(b) Heave, roll and pitch response

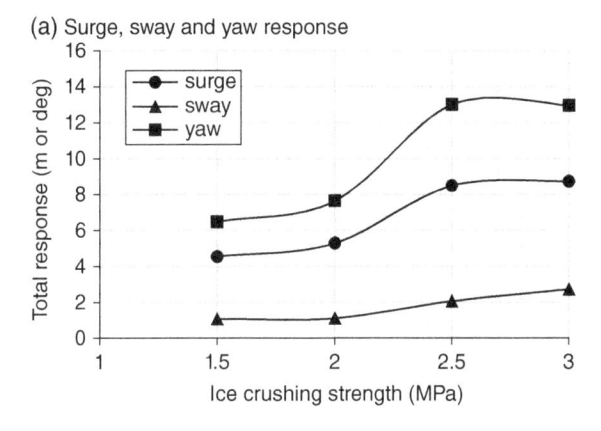

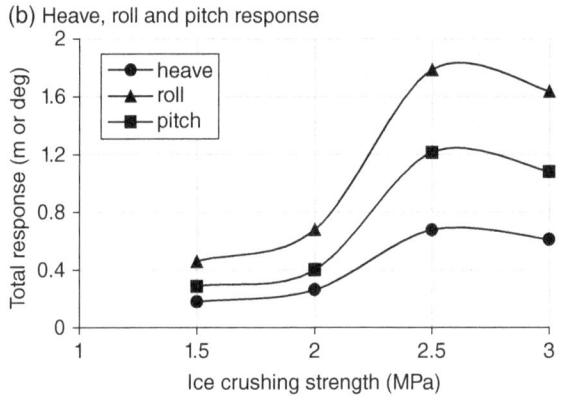

Figure 4.15 Total deck response for different ice crushing strengths.

4.3.3 Comparison of Ice- and Wave-Induced Responses

Ice load actions on two buoyant legs induce a response in all DOF, whereas wave actions result only in the surge, heave, and pitch responses when the load acts in the positive x direction. The surge, heave, and pitch responses of a deck of the triceratops under extreme is sea conditions and high sea conditions in the presence of random wave actions represented by the Pierson-Moskowitz (PM) spectrum are given in Table 4.4. The maximum shift in the surge mean value is observed given the ice load action. However, the maximum response is observed in the open water load action, i.e. in an environment with waves alone. The heave and pitch responses in both cases are found to be very low. The PSD plots of the deck in the surge DOF in both cases are shown in Figure 4.17. The first peak

(a) Surge, sway and yaw response

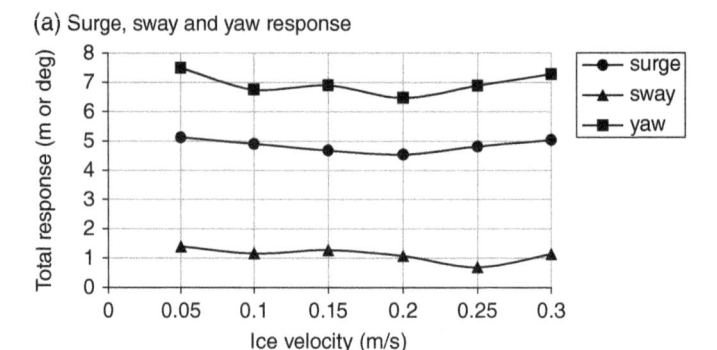

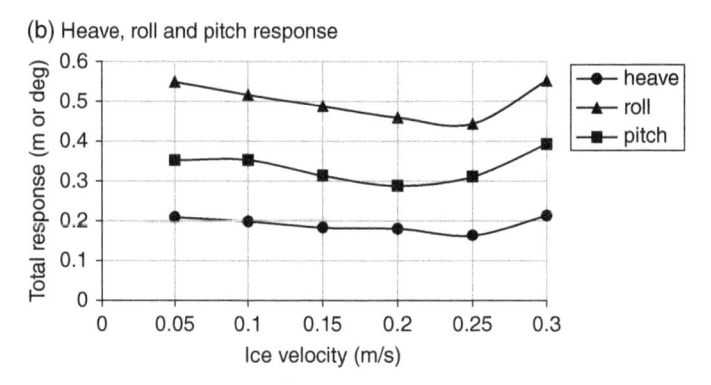

(b) Heave, roll and pitch response

Figure 4.16 Total deck response for different ice velocities.

Table 4.4 Deck response to open water and ice-covered load cases.

Load case	Open water			Ice-covered		
Statistics	Surge (m)	Heave (m)	Pitch (deg)	Surge (m)	Heave (m)	Pitch (deg)
Maximum	6.478	0.0059	0.1145	4.540	0.000	0.047
Minimum	−5.034	−0.1059	−0.0487	0.000	−0.180	−0.241
Mean	0.123	−0.0083	0.0022	2.489	−0.062	−0.077
SD	1.654	0.0117	0.0101	0.902	0.036	0.053
RMS	1.659	0.0144	0.0103	2.648	0.072	0.093

occurs at the natural surge frequency where the response to the open water load case is at a maximum. The second peak occurs at one-fourth of the natural pitch frequency, where the response of the deck to the ice-covered load case is at a maximum. In the open water load case, the mean shift from the initial tether tension is found to be similar in all three buoyant legs. In the ice-covered load case, the

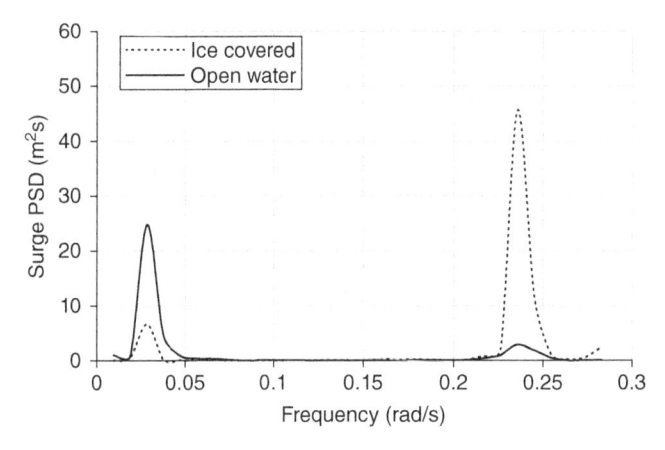

Figure 4.17 PSD plots of the deck in open water and ice-covered load cases.

mean shift is found to be at a maximum in buoyant leg 1. The tether tension variations in the buoyant legs in the open water load case are 8.5, 8.4, and 8.5%. In the ice-covered load case, they are found to be 1.4, 1.2, and 1.4%. Fatigue damage increases from normal to extreme ice sea conditions and hence results in a reduction of the service life. The service life of tethers in normal ice sea conditions is 23.93 years, which decreases to 20.36 years in extreme ice sea conditions.

4.4 Response to Impact Loads

In order to investigate the response of buoyant legs to impact loads, a rectangular box-shaped indenter of length 10.0 m, width 5.0 m, depth 2.0 m, and 7500-ton displacement is considered as a striking mass that represents the stem of a ship, or so-called *stem bar*. The indenter impacts a cylindrical shell at a height of 5.0 m above mean sea level (MSL). The indenter is assumed to be infinitely rigid, and the energy is dissipated only by the platform. A ductility design is followed, which implies that the platform dissipates the major part of the collision energy by undergoing large plastic deformation (Veritas 2010b). Similar to the study of a triceratops under environmental loads, the numerical analysis under different load cases will help in understanding the behavior of the platform. The load cases given in Table 4.5 may be considered in the analysis of offshore structures.

In order to get a clear understanding of the local and global deformation of buoyant legs, the impact analysis is carried out using the methodology shown in Figure 4.18. The buoyant leg is modeled as an orthogonally stiffened cylindrical shell structure using an explicit analysis solver. The indenter is modeled as a rigid solid body without deformation. Thus, the dissipation of strain energy is confined to the shell and the stiffeners. Meshing plays an important role in the numerical

Table 4.5 Collision speed and impact duration (Syngellakis and Balaji 1989).

Impact load case	Collision speed (m/s)	The impact duration (seconds)
Case 1	1.0	0.30
Case 2	2.0	0.35
Case 3	3.0	0.38
Case 4	4.0	0.40

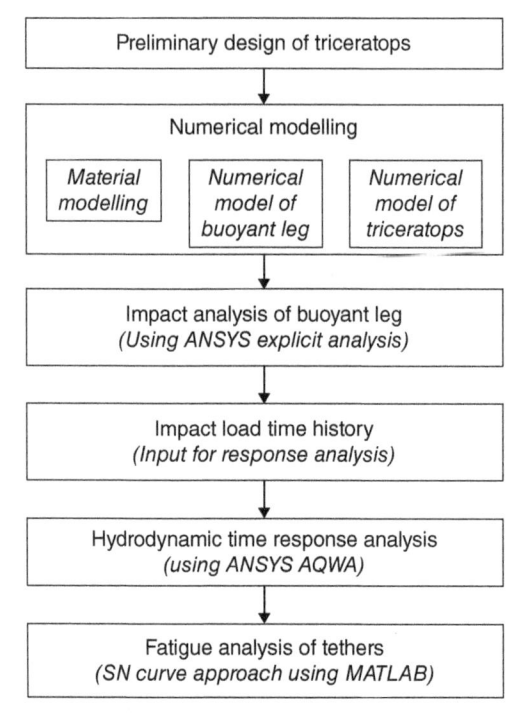

Figure 4.18 Methodology of impact analysis.

analysis. The quality of the meshing should be checked through momentum and energy conservation for different mesh sizes. The striking mass is restrained in all DOF except in the impact direction. An initial collision velocity of 5.0 m/s is applied to the rectangular indenter in the impact direction to simulate impact energy. The developed numerical model is shown in Figure 4.19. The ring stiffeners in the buoyant legs are numbered from R1 to R5 above the MSL. Ring stiffener R3 is located in the impact zone. With due consideration to the height of the visiting vessels in the offshore industry, the indenter is placed 5.0 m above the MSL.

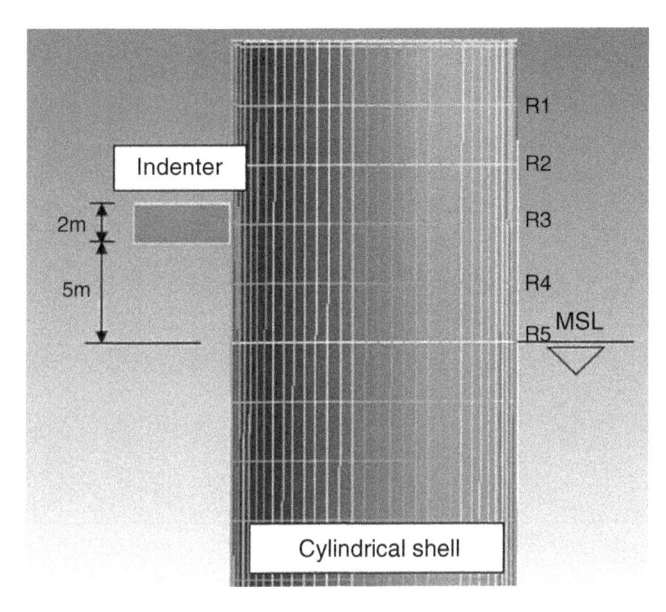

Figure 4.19 Numerical model of buoyant legs and indenters.

The impact load–time history obtained from the analysis is then applied as an external force on the buoyant legs to carry out the hydrodynamic response of the triceratops in the time domain.

The indenter impact causes a local dent, leading to flattening of the outer cylindrical shell of the buoyant leg and ring stiffener at the impact location. The flattening of the local dent increases with an increased contact area between the indenter and the buoyant leg. The ring stiffener hinders the spread of damage to the adjacent bay. Hence, it acts as an obstruction to circumferential bending (Do et al. 2018). The ring frames at the end of the cylindrical shell remain circular, unaffected by impact load. The maximum strain in the cylindrical shell is observed only within the adjacent bays of deformed ring stiffener R3 at the impact location. The stringer stiffeners near the impact location collapsed as a beam between the ring stiffeners. With increased dent depth at the impact location, the stringers adjacent to the damaged stringer also start to deform with the cylindrical shell. Further, local tripping of stringer stiffeners is observed close to the deformed ring stiffener. The maximum equivalent stress and deformation increase in the cylindrical shell with increased impact velocity and duration. The equivalent stress in the buoyant leg increases beyond the yield stress of the material in impact load case 4, which results in the reduction of the load-carrying capacity of the buoyant leg. It may further cause instability in the triceratops. This highlights the adverse effect of high-velocity ship impacts on the survivability of the platform. The force versus nondimensional deformation curves are shown in Figure 4.20. As seen from the figure, the flattening of the curves at a particular instant occurs due to torsional buckling of the stiffeners. It can be clearly said that higher-intensity impact loads may lead to local weakening of the structure.

The impact load–time history obtained from the explicit analysis is applied as an external force in buoyant leg 1. Maximum surge and pitch responses are observed in the impacted buoyant leg. The surge response is transferred from the impacted buoyant leg to the other buoyant legs through the deck. Though the magnitude of the responses is much less, the impact causes continuous periodic vibrations on the deck, as seen in Figure 4.21.

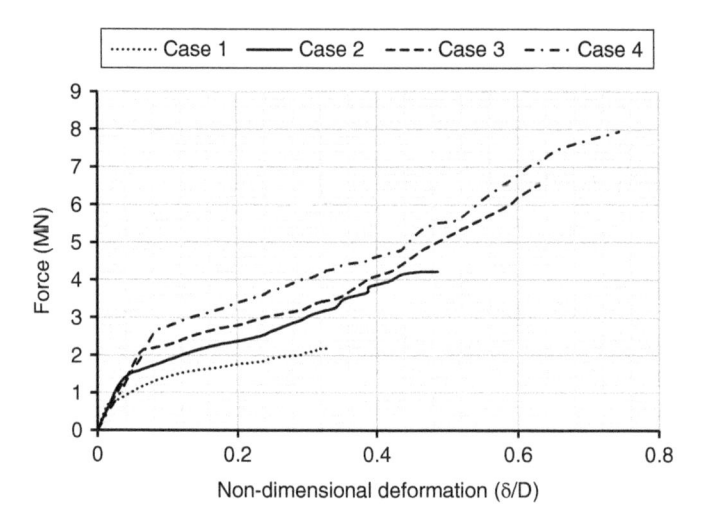

Figure 4.20 Force versus nondimensional deformation curve.

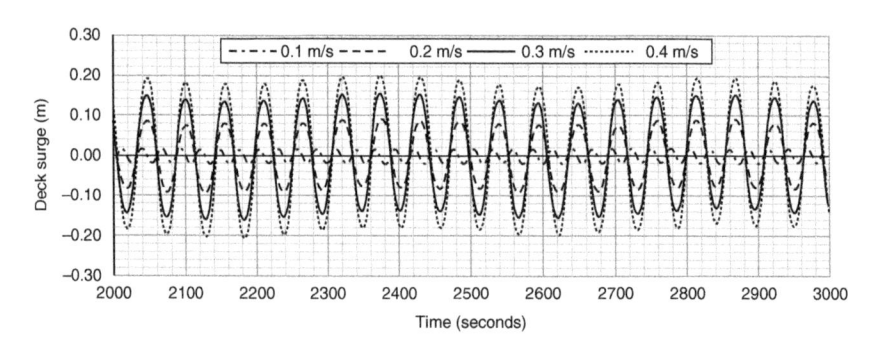

Figure 4.21 Deck surge responses for impact loads on buoyant leg 1.

The surge and pitch responses in the buoyant legs induce significant tether tension variation in the triceratops. The tether tension variation is less than 3% even for the maximum impact load case. The maximum tether tension and the mean shift increase with increased impact velocity. Even though the platform undergoes periodic vibration in different DOF, much less fatigue damage is observed due to fewer stress cycles.

4.4.1 Parametric Studies

4.4.1.1 Indenter Size

The depth of the indenter is varied from 1.0 to 3.0 m. The impact load cases are represented by the b/R ratio, where b is the depth of the indenter and R is the outer radius of the cylindrical shell. The force–deformation curve for different indenter depth cases is shown in Figure 4.22. Greater impact force is developed by the indenter with a lower b/R ratio of 0.133. The elastic spring-back of indenters with a lower b/R ratio occurs faster compared to indenters with a higher b/R ratio, which can be seen from the energy dissipated by the buoyant leg. Indenters with greater depth are more critical and cause maximum deformation in response to comparatively lower impact force, in comparison with indenters with smaller depth. The pattern of damage in the cylindrical shell is not much affected by the size of the indenter. However, the extent of deformation along the length of the cylindrical shell and the bulging of the shell at the ends of the dented region increase with increased depth of the indenter. The surge and pitch responses of the buoyant leg and deck decrease with an increased b/R ratio due to the reduction

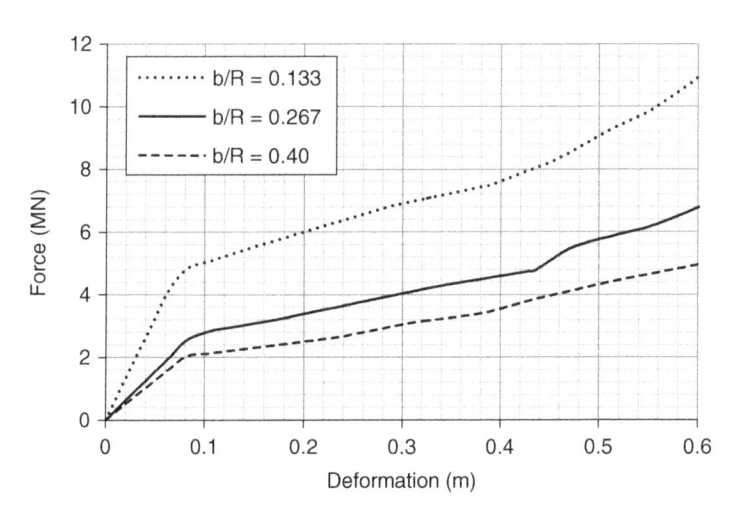

Figure 4.22 Force–deformation curves for different indenter sizes.

in the impact force. Indenters with a reduced *b/R* ratio invoke a concentrated impact force on the buoyant leg, where the peak force occurs within a short time. The short-duration maximum force on the buoyant leg increases the global response of the deck and buoyant legs. Much less surge and heave motion in the triceratops does not induce higher tether tension variation. Hence, fatigue damage in the tethers also is not affected by the change in the size of the indenter (Chandrasekaran and Nagavinothini 2019a, b).

4.4.1.2 Collision Zone Location

The collision zone location is based on the vessel draft, wave height, and maximum tide level. The impact behavior of the structure, especially the local deformation pattern, is highly affected by the impact location. According to DNV regulations, the collision zone may be considered 5.0 m above the MSL. The location of the center of the indenter is varied from 3.0 to 6.0 m above the MSL. Ring stiffeners R3 and R4 are located at impact locations 1 and 5, respectively. The impact load for location 3 acts on the mid-bay between ring stiffeners R3 and R4. The force–deformation curves of the mid-bay buoyant leg for different impact locations for an impact velocity of 4.0 m/s are shown in Figure 4.23.

At impact locations 1 and 5, the ring stiffener undergoes maximum deformation compared to the stringers and cylindrical shell. Though the center of the indenter does not coincide with the location of the ring stiffeners of the buoyant leg at impact locations 2 and 4, the global deformation in the ring stiffeners is comparatively higher than that of cylindrical shell and stringers. At impact locations 2 and

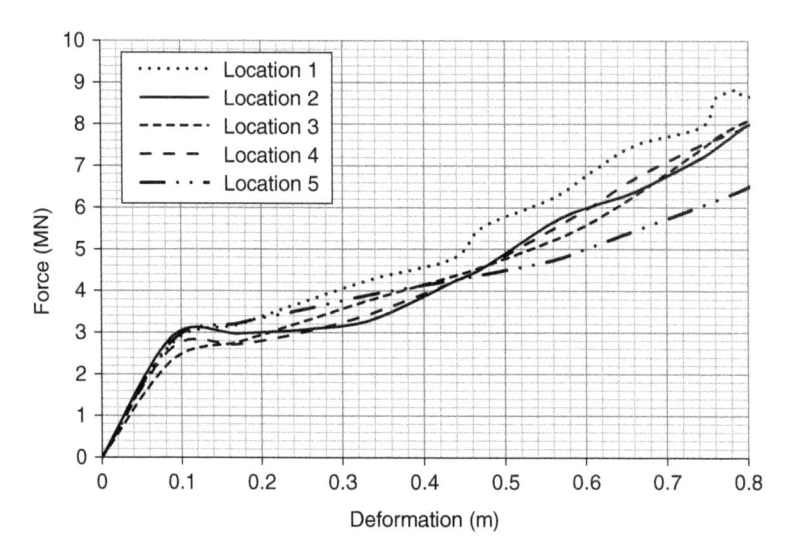

Figure 4.23 Force–deformation curves for different impact locations.

4, yielding occurs initially in the stringers and then is transferred to the ring frames. However, the ring frames play a significant role in resisting the impact force in all cases. At mid-bay impact, the stringers yield like a beam restrained at its ends by the ring stiffeners. Given this condition, the stringers yield first and then transfer the loading to the ring frames, leading to increased global deformation of the shell. Due to less variation in the impact force developed at the different locations of the indenter, not much variation is seen in the response of the deck and buoyant legs of the triceratops. The rate of response decay is almost the same for all cases. The ball joint effectively reduces the transfer of pitch motion from the buoyant legs to the deck. The pitch response in the deck is reduced by up to 99% of the pitch response of the buoyant legs. Fatigue damage in all cases remains the same as 8e-6 due to similar tether tension variation in the tethers in all the cases.

4.4.1.3 Indenter Shape

The force–deformation curves of the knife-edge indenter and hemispherical indenter are shown in Figure 4.24. As can be seen from the pattern of the force–deformation curves, flattening of the curve is greater under the impact of the knife-edge indenter. The knife-edge indenter also causes maximum deformation in the stringers. Due to the reduced cross-sectional area, the hemispherical indenter does not cause flattening of the ring stiffener and cylindrical shell. It causes maximum deformation in the ring stiffener. For all types of indenters, the deformation starts by flattening the circumferential curvature, and it also leads to compressive membrane strains. The cylindrical shell also shows higher bulging at the ends of the flattened section. This confirms that indenters with pointed ends

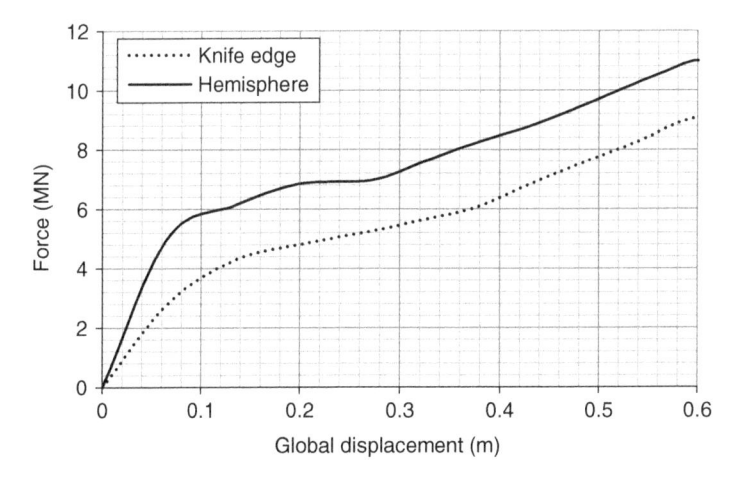

Figure 4.24 Force–deformation curves for different indenter shapes.

may cause severe local damage to the cylindrical shell. The reduced contact area of hemispherical indenters reduces the flattening of the cylindrical shell but distorts the shell by increasing the circumferential strain. The stiffeners in the impact location are heavily damaged in this case. In comparison, flattening of the ring stiffener is significant in response to the impact of a knife-edged striking mass.

4.4.1.4 Number of Stringers

The change in the number of stringers has only a small effect on the impact response characteristics of the buoyant legs, as seen in Figure 4.25. The peak force increases by less than 1% with the increase in the number of stringers for the same deformation. It occurs due to increased resistance offered by the stringers to the indenter. Due to much less variation in the peak force with time, the variation in the deck response of the triceratops is also negligible. The closely spaced stiffeners reduce the maximum deformation marginally. It can be seen that the increase in the number of stringer stiffeners does not provide excessive resistance to external loads. This should be taken into account when determining the number of stringers during the design of the buoyant legs, to achieve weight efficiency and optimum impact resistance.

4.4.2 Impact Response in the Arctic Region

Ship impact collisions in the Arctic region pose an additional threat to offshore structures due to the low prevailing temperatures. The lowest temperature in the Arctic islands and continental regions during winter can be below −60 °C. The

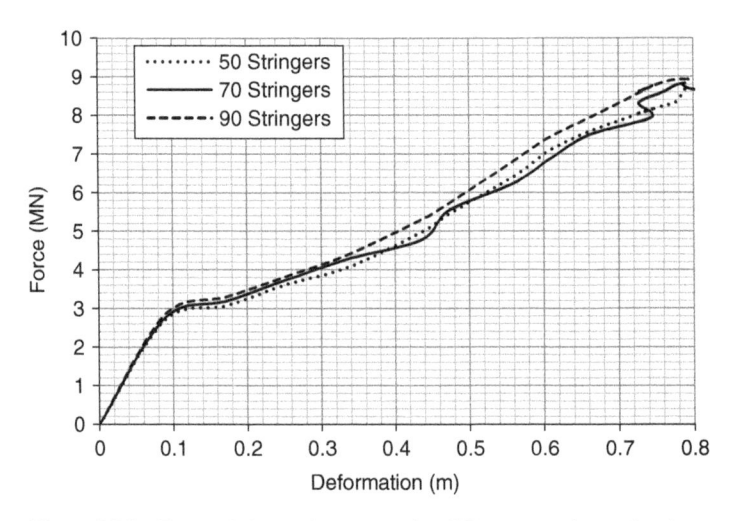

Figure 4.25 Force–deformation curves for different numbers of stringers.

structural steel used in the construction of offshore platforms suffers from reduced toughness at such low temperatures, which in turn affects the performance of the platform. Steel used in an Arctic environment must satisfy fracture toughness requirements at temperatures between −40 and −60 °C. The steel's toughness can be increased by decreasing the grain size of the steel and adding magnesium, copper, chrome, and nickel (Jumppanen 1984). An example of one such material is DH36 polar-class high-tensile steel. The mechanical properties of DH36 steel are given in Table 4.6.

An impact analysis can be carried out at Arctic temperatures using the methodology explained previously. The force–deformation curves at room temperature and Arctic temperatures are shown in Figure 4.26. The flattening of the curve at 0.1 m indicates the torsional buckling of the stiffeners. In both cases, the force–deformation curves are almost identical in the initial stages, showing good resistance to deformation. However, the ultimate resistance to deformation is reduced at Arctic temperatures irrespective of the increased stiffness in the cylindrical shell. This can be attributed to the reduction in fracture toughness of the material.

Table 4.6 Mechanical properties of DH36 steel at a 0.001/s strain rate (Kim et al. 2016).

Temperature (°C)	Yield Strength (N/mm^2)	Ultimate strength (N/mm^2)
Room temperature (RT)	383.7	530.2
Arctic temperature (AT)	446.2	606.5

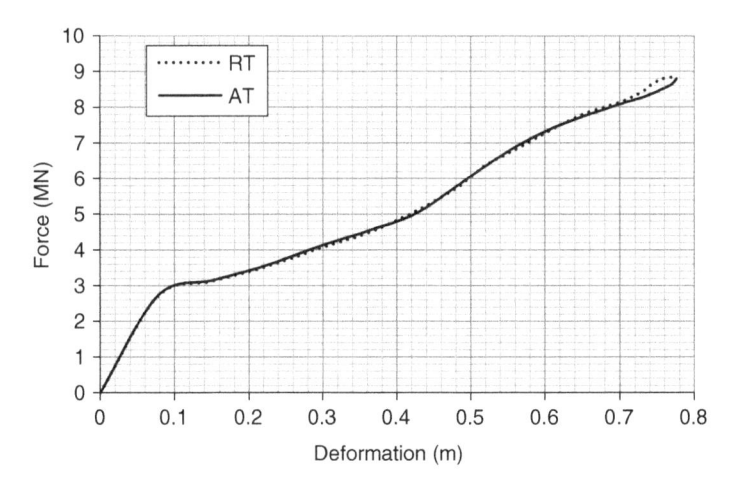

Figure 4.26 Force–deformation curve of buoyant legs at different temperatures.

4.5 Deck Response to Hydrocarbon Fires

The deck plates of offshore structures are usually designed as stiffened plates. The behavior of stiffened plates in response to a hydrocarbon fire depends mainly on the type of material, mechanical loads on the structure, the means of heating, and the maximum temperature attained. The behaviors of the internal and external elements vary during free actions. Internal elements receive thermal loads that are comparable to those obtained from standard fire tests. However, external elements are subjected to radiation from the surface, heat from windows, and convection from flames, which changes their characteristics significantly. It is vital to assign the material properties to the numerical model in order to assess the exact behavior of the structure under consideration. Normally, the material properties will be defined using multilinear stress–strain curves at different temperatures.

In the case of the offshore triceratops, a typical deck plate is 6.0–9.0 m long and 12 mm thick. The plates are supported by intermediate stiffeners and girders, as shown in Figure 4.27. In order to reduce the runtime, the analysis can be carried out on a scale model of the deck plate. The scale plated model measuring 1.375 m × 1.250 m × 0.003 m is considered, as shown in Figure 4.28 (Gruben et al. 2016). The L-shaped stiffeners are placed 0.2 m apart and are numbered S1 to S6. The plate is modeled as a shell element and is considered to be fixed on all four edges.

The increase in temperature in the deck plate and the thermal load that develops on a structure in response to a particular fire condition can be predicted using thermal analysis, which is carried out using a standard time–temperature curve. The location of the outbreak of a hydrocarbon fire in an offshore platform depends upon several conditions, and it varied in all the accidents reported earlier. Thus, a

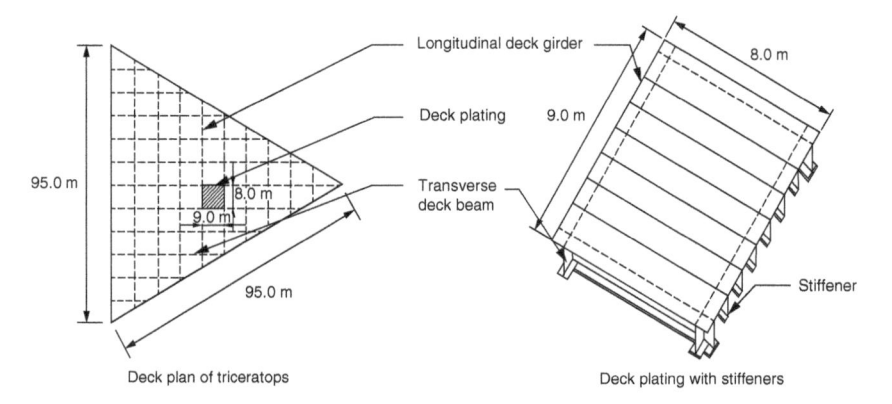

Figure 4.27 Deck plate of a triceratops.

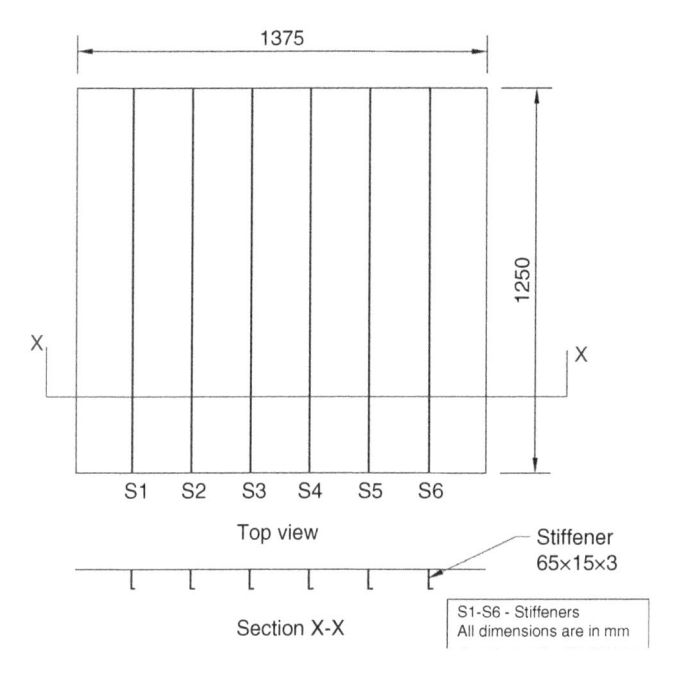

Figure 4.28 Scale deck plate model.

thermal analysis should be carried out for different fire cases, similar to the different load cases considered for analysis under environmental and impact loads. The fire analysis of deck plates can be carried out by considering the following fire scenarios, where the fire is assumed to be concentrated on the edge bay, mid-bay, and whole plate, as shown in Figure 4.29.

The hydrocarbon fire acts on the top of the deck plate, and the fire is transferred to the stiffeners by conduction. The temperature increases in the deck plate and stiffeners are shown in Figure 4.30. In case 1 (an edge fire), the temperature of the plate increases to 606 °C at the bay where the fire is concentrated. The temperature increases to more than 400 °C for 13% of the total area of the plate (near the edge). In case 2, the heat load is transferred in both directions along the length of the plate, which subsequently reduces the maximum temperature of the plate compared to the earlier case; the temperature of about 13% of the total area of the plate is increased beyond 400 °C due to the fire. Case 3, where the entire top face of the plate is engulfed in fire, shows the maximum temperature of the plate as 687.25 °C, which is higher than the previous cases. As seen from the temperature distribution, the constraints at the end of the plate increase the temperature at the ends of the plate and stiffeners 1 and 6 (S1 and S6).

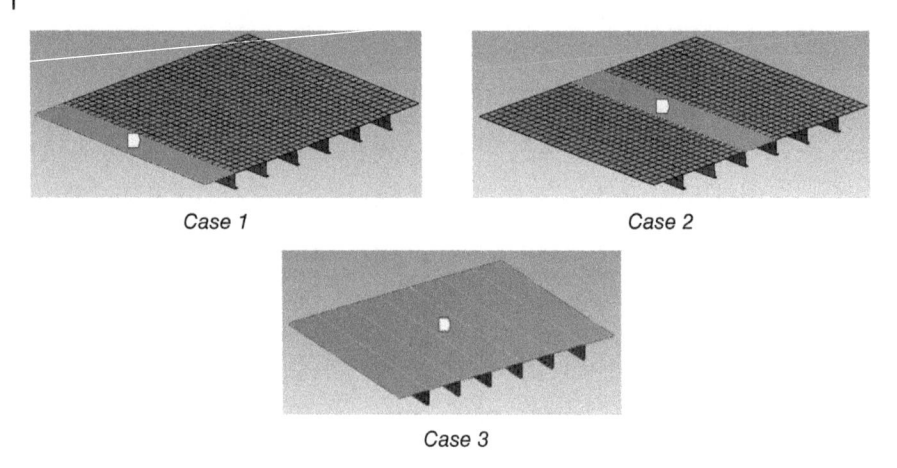

Figure 4.29 Hydrocarbon fire cases.

The thermal load developed in all these cases induces axial stresses on the plate and degrades the material properties. Increased thermal stress on the plate results in the bending of the plate, which in turn transfers initial compression to the stiffeners. This causes more deformation in the stiffeners and reduces the strength of the stiffened plate considerably. In the case of the hydrocarbon fire acting on the edge of the plate, the thermal stress developed in the plate exceeds the yield stress at 1100 seconds. The plastic zone was first developed at the outer edge of the plate, which was subsequently transferred to the mid-span. Thus, the fire rating of the unprotected stiffened steel plate given an edge-fire condition is about 18 minutes from the onset of the fire and reaches a maximum temperature of about 315 °C. The fire rating for the mid-bay fire scenario is around 20 minutes, which is slightly more than that of the edge-fire condition. The reduced fire rating for the edge-fire scenario can be attributed to the edge support close to the bay, which restrains the distribution of stresses. In the third case, the thermal stress increases simultaneously over the entire area of the plate and reaches the yield stress at 650 seconds, causing maximum deformation at 1000 seconds. As the entire top face of the plate is engulfed in fire, the fire rating is reduced to 10 minutes in the third fire scenario.

4.6 Summary

Offshore structures need to be analyzed for dynamic effects arising due to ice–structure interactions. Of several ice failure modes, crushing ice failure tends to cause maximum ice force on the structure. In the case of multi-legged structures

Triceratops Under Special Loads | 159

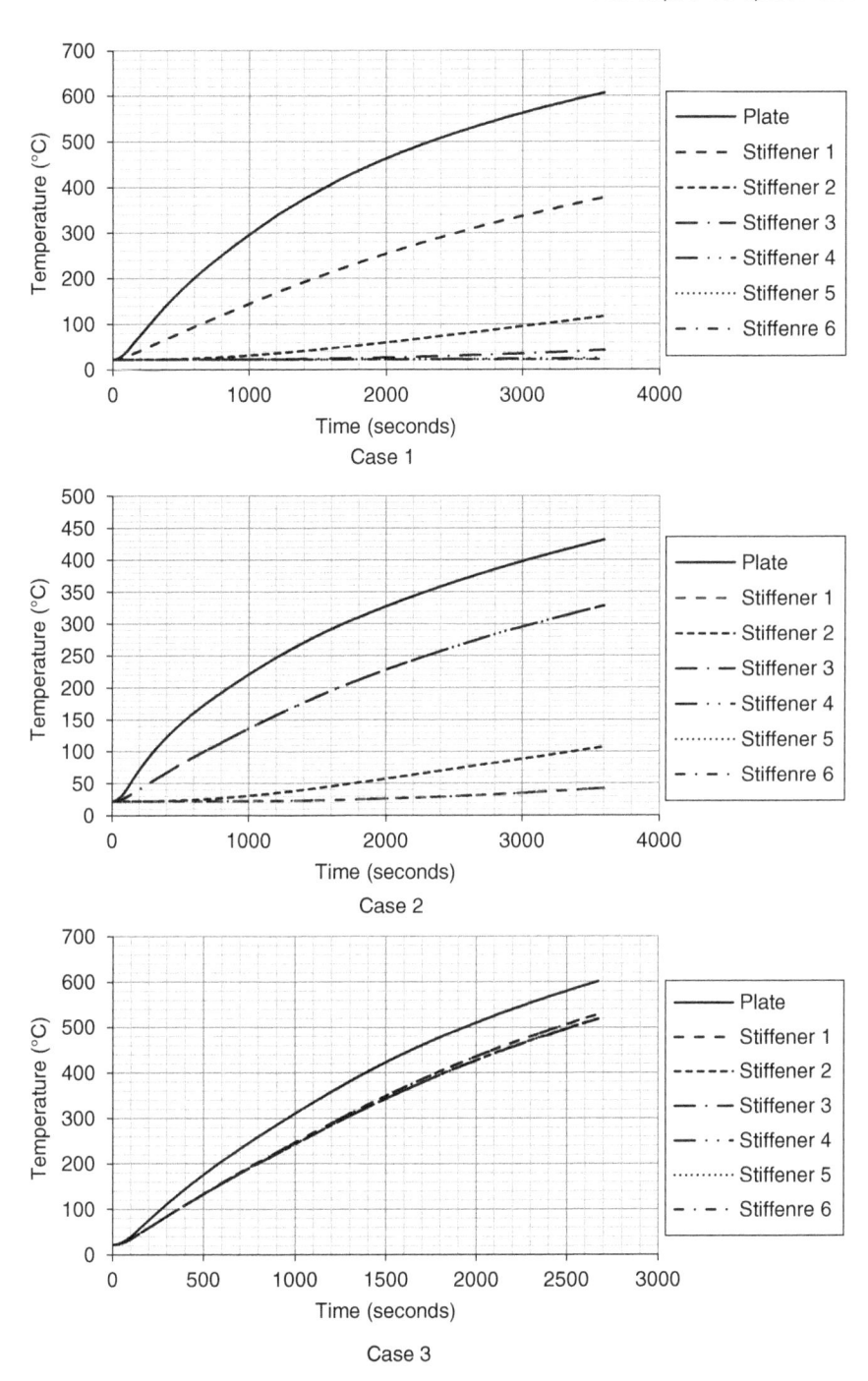

Figure 4.30 Temperature variations in plates and stiffeners.

like the triceratops, the maximum ice load condition develops when ice hits two legs simultaneously. The continuous crushing of the ice on two buoyant legs causes a response in all DOF. The compliant DOF of the platform is highly affected by the action of ice forces. The ball joints restrain the transfer of rotational motion from the buoyant legs to the deck. With increased ice thickness and ice crushing strength, the total response in all DOF increases, along with an increased mean shift in both the deck and buoyant legs of the triceratops. The fatigue analysis results show an increase in fatigue damage and a decrease in the service life of tethers from normal to extreme ice sea conditions.

The buoyant legs of the triceratops are prone to accidental impact loads arising from ship–platform collisions. The impact load causes a local dent, leading to flattening of the cylindrical shell and ring stiffener at the impact location. The flattening of the local dent increases with an increased contact area of the indenter. The ring stiffeners prevent the damage from spreading to the adjacent bay and act as obstructions to circumferential bending. The deformation in the buoyant legs increases with increased impact velocity and duration. The maximum deformation increases with increased distance of the indenter from the top end of the cylinder. The dent depth increases with an increased b/R ratio due to the increased area of contact. This also significantly increases the response of the deck and the buoyant legs. The spread of damage in the longitudinal direction is comparatively less for a hemispherical indenter than for a knife-edge indenter, where the damage is highly concentrated on the impact location. The change in the number of stringers has only a minor effect on the impact response characteristics of the buoyant legs.

Hydrocarbon fires are among the most severe accidental loads on topside structures in the offshore industry. The behavior of the plate is influenced by a few factors: the location and duration of the fire, the area of the plate under the direct influence of the fire, and the additional load imposed on the plate. The thermal stress developed in the plate exceeds the yield stress of the material in less than 10 minutes when the entire top face of the plate is covered by fire. Increased thermal stress on the plate results in the bending of the plate, which in turn transfers initial compression to the stiffeners. This causes more deformation of the stiffeners and reduces the strength of the stiffened plate considerably.

5

Offshore Triceratops

Recent Advanced Applications

Summary

This chapter explains the application of offshore triceratops as support systems for wind turbines. The response behavior of an offshore wind turbine supported on a triceratops, obtained through numerical investigations, is also presented. Fatigue analysis and service life prediction of tethers are also discussed. This chapter also presents recent developments in conceptual models of triceratops with different types of buoyant leg configurations. Conceptual models of stiffened and elliptical buoyant legs are presented. The detailed response of the triceratops to wave actions is also presented.

5.1 Introduction

One of the most sensitive studies carried out in recent years is the assessment of the motion characteristics of any offshore structure. Because a triceratops is one of the most advantageous new-generation offshore compliant structures, attempts have been made to improve the conceptual model of the platform. In addition to geometric innovations made to the platform with ball joints, the configuration of the buoyant legs can be conveniently modified to improve the platform's advantages and its suitability in adverse environmental conditions.

5.2 Wind Turbines

Serious challenges associated with the extraction of primary sources of energy lead to a need for alternative sources of energy that are better options for energy extraction. Wind is an inexhaustible energy source and thus is one of the possible

Offshore Compliant Platforms: Analysis, Design, and Experimental Studies,
First Edition. Srinivasan Chandrasekaran and R. Nagavinothini.
© 2020 John Wiley & Sons Ltd.
This Work is a co-publication between John Wiley & Sons Ltd and ASME Press.

options for energy extraction with no adverse threat to the environment. Due to the unavailability of land, offshore wind energy extraction is becoming increasingly attractive. Offshore wind turbines with capacities greater than 900 MW have been installed in the Baltic Sea and North Sea (Musial et al. 2006). Compared to onshore wind farms, offshore wind farms are considered to have greater potential due to the consistency and strength of the wind. The major advantages of offshore wind turbines are as follows:

- Less intense sea turbulence
- Fewer constraints on the size of the wind turbines
- Avoidance of noise and visual disturbances due to their distance from shore

Though the vast, uninterrupted open sea is available for wind energy extraction without interference from land uses, there are also some disadvantages associated with the installation of wind turbines at sea:

- Very high initial investment
- Complications involving the construction of the foundation and supporting structure, commissioning, and decommissioning
- Less accessibility compared to onshore wind farms, which in turn increases downtime and increases the cost of maintenance and operation
- Complexities arising due to the extreme hydrodynamic and aerodynamic loads acting on the supporting structures and turbines

Fixed structures are widely used as supporting structures for wind turbines in shallow water depths of less than 20 m. However, these structures are highly expensive in deep water. Floating platforms such as the spar and tension leg platform (TLP) are widely used in deep water to support wind turbines. These floating structures are technically feasible: they have been used for years by the offshore industry for oil and gas production. Other technological and economic challenges that arise when replicating offshore technology for wind farms should be addressed through proper design and thorough conceptual analyses (Butterfield et al. 2005).

It is also mandatory to follow standard guidelines during the design of wind turbines. The design requirements for land-based wind turbines are given by the International Electrotechnical Commission (IEC) 61400-1 design standard (IEC 61400-1 2005). The IEC 61400-3 design standard acts as an attachment to the IEC 61400-1 design standards for sea-based wind turbines.

Offshore wind turbines require integrated load analyses with comprehensive simulation tools. One of the major complexities in the design of floating offshore wind turbines (FOWTs) is the different combinations of loading. The most common loadings on offshore wind turbines include wave- and platform-induced hydrodynamic loads, wind loads, impact loads from floating debris, sea ice, and marine growth on the substructure. In addition, the dynamic coupling motion between the platform and turbine should be considered.

5.3 Wind Power

In the last few years, a large number of offshore wind farms have been put into operation in European countries such as Denmark, the United Kingdom (UK), and the Netherlands in shallow water (less than 25 m), relatively close to the shore. Simple concrete gravity structures and steel monopiles have proven to be economical for such developments (Musial and Butterfield 2006). The major offshore wind projects in terms of installed power are as follows:

- Lynn and Inner Dowsing (194 MW), UK
- Kentish Flats (90 MW), UK
- Burbo Bank (90 MW), UK
- Princess Amalia (120 MW), Netherlands
- Nysted (165 MW), Denmark
- Horns Rev (160 MW), Denmark

New offshore wind energy projects are being carried out in deep water. For example, one recent offshore project is the 2 MW offshore wind turbine installed in northern Portugal on a floating device called WindFloat. The cost of the substructure and the foundation is higher than it is for wind turbines installed in shallow water, so an economically feasible design is necessary for overall project viability.

5.4 Evolution of Wind Turbines

In the early and mid-1980s, the capacity of a typical wind turbine was less than 100 kW. By the late 1980s and early 1990s, turbine capacity had increased from 100 to 500 kW. Further, in the mid-1990s, typical capacity ranged from 750 to 1000 kW. And by the late 1990s, the installed turbine capacity had increased up to 2.5 MW. Now turbines are available with capacities above 5 MW. Based on water depth, offshore wind turbines are classified as follows:

- Shallow-water wind turbines (commissioned between 5 and 30 m water depth)
- Transitional wind turbines (commissioned between 30 and 60 m water depth)
- FOWTs (commissioned at water depth greater than 60 m)

Wind turbines in shallow waters generally rest on monopile, gravity-base, or suction-bucket structures, while transitional turbines are supported on a tripod tower, guyed monopile, full-height jacket, submerged jacket, or enhanced suction-type structure. FOWTs are supported by compliant offshore structures like spar platforms, TLP, semi-submersible platforms, or pontoon-type systems.

With the recent proposal of offshore wind power projects in deep water to capture higher-velocity wind, FOWTs resting on compliant offshore platforms are under wide exploration. An attempt is made here to examine the response of an offshore triceratops as a supporting structure for a wind turbine. In addition to reducing risk and minimizing lifecycle costs, the top deck of the triceratops that supports the wind turbine remains isolated from the buoyant legs by ball joints.

5.5 Conceptual Development of the Triceratops-Based Wind Turbine

The two distinct types of wind turbines are the horizontal axis wind turbine (HAWT) and the vertical axis wind turbine (VAWT). The seven major subsystems in the wind turbine are as follows:

- Blades
- Nacelle
- Controller
- Generator
- Rotor
- Tower
- Floating body

The rotor houses a number of blades that determine the system performance of the wind turbine. A three-bladed upwind design is predominantly used in the design of the rotor, and the blade design is usually based on the pitch control. The nacelle protects the generator, controller, gearbox, and shafts. The tower supports the wind turbine nacelle and rotor. The total height of the tower at the particular site is usually governed by the rotor diameter and the nature of the loading conditions. Generators are used to convert the raw mechanical work of the wind turbine to useful electrical output. Changes in the blade pitch angle, generator loading, and nacelle yaw are monitored by the control system. The generated electrical output is transferred to a suitable electrical grid through cables buried in the seabed. If this method is uneconomical, in recent years the generated power has been transferred through battery storage. The wind turbine elements are supported on a floating body, and mooring systems are usually employed to position restrain the system.

5.6 Support Systems for Wind Turbines

Five major categories of FOWT concepts are at various stages of development:

- Spar
- TLP
- Pontoon (barge)

- Semi-submersible
- Triceratops

5.6.1 Spar Type

This type of platform consists of a supporting foundation known as the *floater*, the tower, and the rotor nacelle assembly (RNA). The supporting foundation typically consists of a steel/concrete cylinder filled with gravel and water to keep the center of gravity below the center of buoyancy, which ensures that the entire superstructure stays upright owing to the large righting moment arm and greater inertial resistance to pitch and roll motions. It is not easy for a spar platform to capsize, owing to the large draft of the floating foundation, which is greater than or equal to the hub height to reduce the heave motion and increase the stability of the platform.

A spar type of FOWT must be deployed in deep water, as sufficient keel-to-seabed vertical clearance is required in order for the mooring system to be effective. Such a wind turbine is positioned using a catenary-based mooring system employing anchor chains, steel cables, or synthetic fiber ropes. The turbine is towed in a horizontal position in calm waters close to the deployment site; this is followed by upending and stabilizing the turbine, after which a derrick crane barge is used to mount the tower.

5.6.2 TLP Type

A TLP FOWT consists of a supporting platform structure to carry the wind turbine. A smaller version of the conventional hull form is the mini-TLP, which has been adopted for TLP FOWTs. The commissioning and assembly of the TLP can be carried out onshore, greatly reducing the difficulties encountered in assembling it offshore. Vertical tendons, also called *tethers*, position-restrain the system. The tethers are anchored with a template foundation, suction caissons, or piles. In comparison with other floating structures such as a spar, semi-submersible, or pontoon, a TLP has relatively less dynamic response to wave excitation (Matha 2009).

5.6.3 Pontoon (Barge) Type

A barge-type FOWT uses a large pontoon structure to carry the wind turbine. The concepts of distributed buoyancy and a weighted water-plane area are employed to achieve the required stability and righting moment. The barge is typically moored using catenary anchor chains. The primary disadvantage of this type of FOWT is its vulnerability to the roll and pitch motions generated by oceangoing ship-shaped vessels. Hence, its suitability is limited to calm seas such as those of a lagoon or a harbor.

5.6.4 Semi-Submersible Type

A semi-submersible platform consists of large columnar tubes that are connected to each other through tubular members. Two primary possibilities for the design of semi-submersible FOWTs are as follows: (i) a wind turbine is placed on one of the columnar tubes; or (ii) a cluster of wind turbines sits on all the columns. The water-plane area of the columns provides stability to the system under floating conditions, while the shallow draft expands the number of sites where such a system can be employed (Mahilvahanan and Selvam 2010).

5.6.5 Triceratops Type

In the recent past, there has been an evolution in the geometric forms of offshore platforms in terms of both innovativeness and motion characteristics for deepwater and ultra-deepwater applications. Triceratops platforms are found to be more suitable for deep and ultra-deep water (White et al. 2005). Novelty in the conceptual design of such platforms through the incorporation of ball joints between the deck and buoyant legs makes it distinct from other new-generation offshore platforms.

5.7 Wind Turbine on a Triceratops

The properties of a triceratops-based wind turbine at a water depth of 600 m are listed in Table 5.1. A numerical model of a triceratops with a wind turbine mast is shown in Figure 5.1. The wind turbine is considered a point mass on the deck for the detailed analyses of the structure.

5.8 Response of a Triceratops-Based Wind Turbine to Waves

5.8.1 Free-Decay Response

In the presence of the action of regular waves, the ball joints are effective in isolating the deck of the triceratops from the buoyant legs. This is evident from the pitch response amplitude operator (RAO) plots of the deck and buoyant legs shown in Figure 5.2. The presence of ball joints prevents 70% of the pitch response and 95% of the roll response from being transferred from the buoyant legs to the deck. This improves the overall stability of the platform. The ball joints transfer translational motion such as surge, sway, and heave from the buoyant legs to the deck under the action of hydrodynamic loads.

Table 5.1 Properties of the triceratops-based wind turbine.

Description	Units	Values
Water depth	m	600
Density of material	kg/m^3	7850
Center to center distance	m	70
Diameter of each leg	m	17
I_{xx}, I_{yy} (buoyant leg)	Ton-m^2	26 016 366
I_{zz} (buoyant leg)	Ton-m^2	688 795.11
r_{xx}, r_{yy} (buoyant leg)	m	51.46
r_{zz} (buoyant leg)	m	8.37
I_{XX}, I_{YY} (deck)	Ton-m^2	3 558 415.8
I_{ZZ} (deck)	Ton-m^2	5 894 093.2
r_{xx}, r_{yy} (deck)	m	29.6
r_{zz} (deck)	m	38.1
Area of deck	m^2	1732.4
Freeboard	m	25
Draft	m	76.7
Metacentric height	m	12.246 33
Stiffness of tether	kN/m	84 000
Buoyancy	kN	86 740
Self-weight of the triceratops	kN	39 818.79
Weight of the wind tower mast	kN	6842
Initial tether tension	kN	46 921.21

The natural frequencies of the triceratops given different degrees of freedom (DOF) are assessed through power spectral density (PSD) plots. The surge PSD plot is shown in Figure 5.3. The peak in the surge DOF occurs at 0.016 Hz, which is the surge natural frequency of the triceratops. Similarly, peaks in the heave, roll, and yaw DOF occur at 0.438, 0.239, and 0.025 Hz, respectively. The natural frequencies of the surge and sway and those of the roll and pitch remain the same due to the symmetry of the platform. The PSD plots shift to the right of the wave spectrum under consideration in the case of the stiff DOF and to the left in the flexible DOF. This can be seen from the free-decay PSD plot in the roll DOF shown in Figure 5.4. The peaks of the wave spectrum and rotor frequency are separated from the peaks of the other responses by a safe margin. Hence, it can be concluded that wave- and wind-driven rotor excitations fail to excite the system in any DOF,

Figure 5.1 Numerical model of a triceratops with a wind turbine.

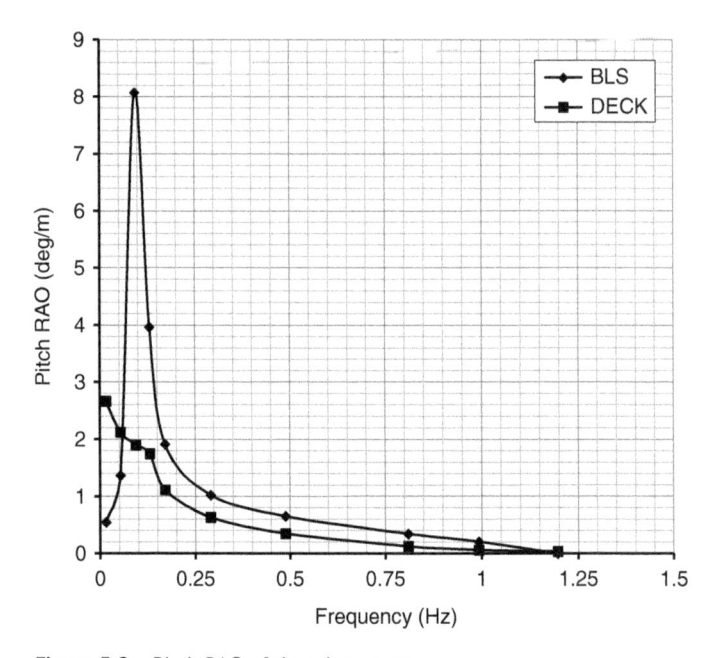

Figure 5.2 Pitch RAO of the triceratops.

making the system dynamically safe and stable. The area of overlap between the wave spectrum and response spectrum is about 31.50, 23.85, 36.10, and 38.80% in the surge, heave, pitch, and yaw DOF, respectively. Though the heave natural frequency lies very close to the rotor frequency, the deck is not subjected to string

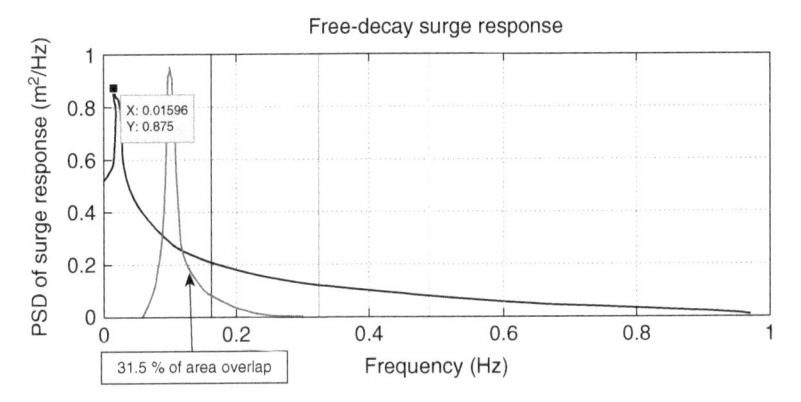

Figure 5.3 PSD plot of the surge free-decay response.

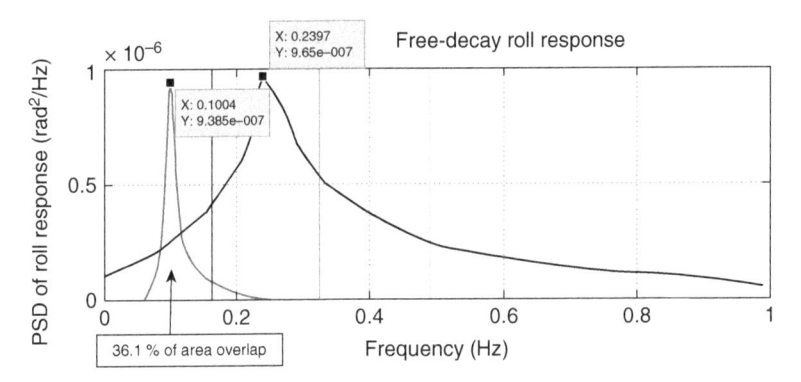

Figure 5.4 PSD plot of the roll free-decay response.

excitations due to the monolithic action of the deck and buoyant legs of the triceratops and the increased stiffness in the vertical plane.

5.8.2 Response to Operable and Parked Conditions

The frequency responses of the deck to the surge, heave, and pitch DOF are shown in Figure 5.5. The magnitude of the response in the flexible DOF is similar in both the operating and parked conditions. The heave peak magnitude is also similar because the operating mass of the entire wind turbine is about 17% of the total mass of the triceratops. The increase in the stiffness of the system is relatively high compared to the mass of the system. Thus, the natural frequency in the heave DOF remains unaltered. A distinct shift in the frequencies occurs in the case of the roll and pitch responses to the turbine's operable and parked conditions. This can be attributed to the coupling between the operating wind turbine and the platform.

170 | *Offshore Compliant Platforms*

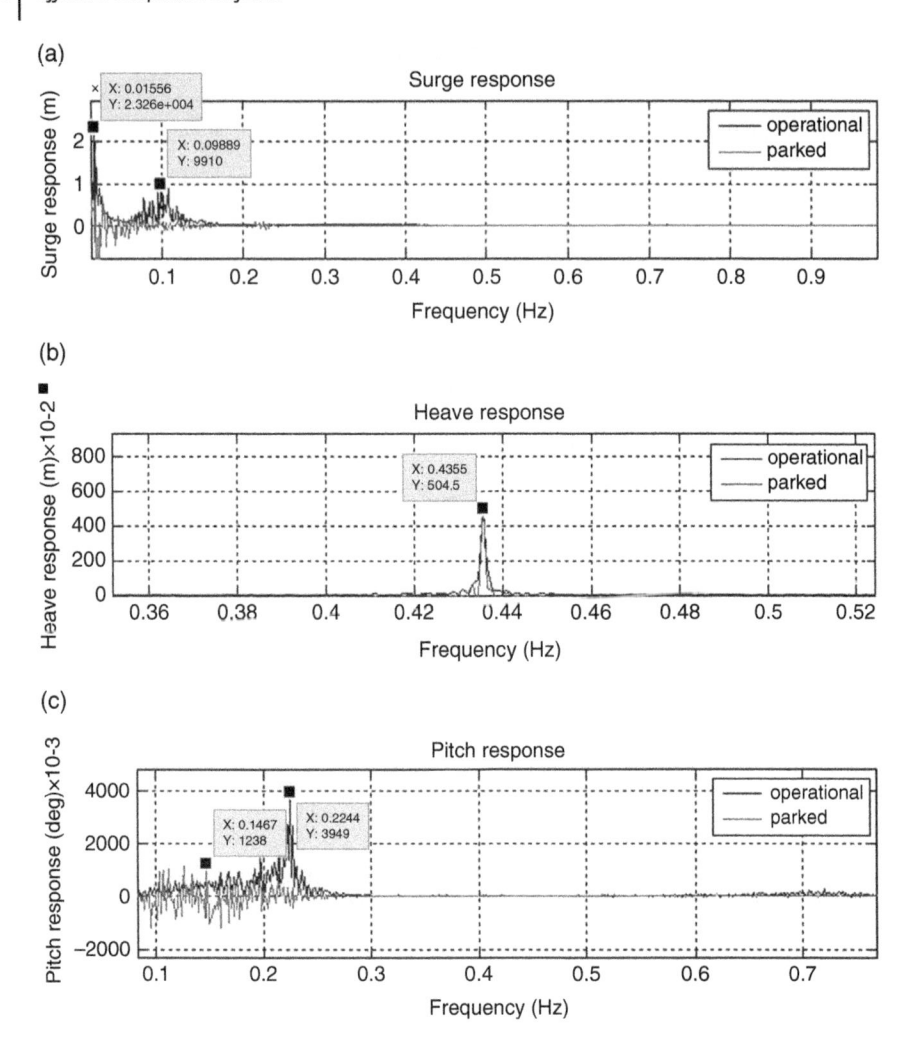

Figure 5.5 Frequency response to operable and parked conditions.

5.8.3 Effect of Wave Heading Angles

The RAOs in different DOF given different wave heading angles are shown in Table 5.2. The direction of the wind is assumed to be the opposite of the 0° wave direction. The surge response decreases and the sway response increases with an increase in the wave heading angle from 0 to 90°, due to the corresponding decrease and increase in the relative velocities between the waves and wind. The surge response at 0° is not equal to the sway response at 90° due

Table 5.2 Variation in RAO with changes in the wave heading angle.

Wave heading	Surge (m)	Sway (m)	Heave (m)	Roll (deg/m)	Pitch (deg/m)	Yaw (deg/m)
0	0.857	0.012	5.03E-03	0.029	0.036	7.71E-03
30	0.634	0.272	1.54E-02	0.030	0.034	7.36E-03
45	0.496	0.502	1.72E-02	0.032	0.032	3.25E-02
60	0.254	0.711	3.06E-02	0.032	0.030	5.51E-02
90	0.012	0.852	3.67E-02	0.035	0.027	6.85E-02

to the reduced aerodynamic damping in the surge response compared to the sway response. The roll response increases with an increased wave heading angle due to the coupling between the roll and sway DOF. Similarly, the coupling between the surge and pitch DOF causes an increase in the pitch response with an increased wave heading angle. Due to the symmetry of the platform, the yaw response remains the same given different wave heading angles. However, the combined surge and sway actions along with aerodynamic thrust increase the yaw response at 90°.

5.8.4 PSD Plots

The PSD plots for different DOF are shown in Figure 5.6. In the surge response, the peak occurs close to the surge natural frequency and peak wave frequency, and low-frequency excitations occur due to wind turbulence. As the wave direction is along the positive x-axis, the standard deviation of the sway response is 22% less than that of the surge response. In the heave response, the peak occurs at the wave frequency and also at the surge natural frequency due to the coupling between the surge and heave DOF, which is inherent in compliant platforms. In the roll and pitch responses, peaks occur at the respective natural frequencies. As the roll natural frequency is very close to the peak wave frequency, it results in a low-frequency beating effect. Low-frequency excitations are driven by turbulence in both the roll and pitch responses. The effect of all the DOF can be seen in the yaw response through the smaller peaks at different natural frequencies of the system. Rotor load excitation is seen as a peak at 0.49 Hz, suggesting the influence of rotor loads on the yaw response. The yaw PSD is highly noisy compared to all the other responses. The yaw response is never a function of direct loading; it is primarily caused by rotation of the mooring lines, which are under the influence of the platform's combined roll and pitch responses.

172 | *Offshore Compliant Platforms*

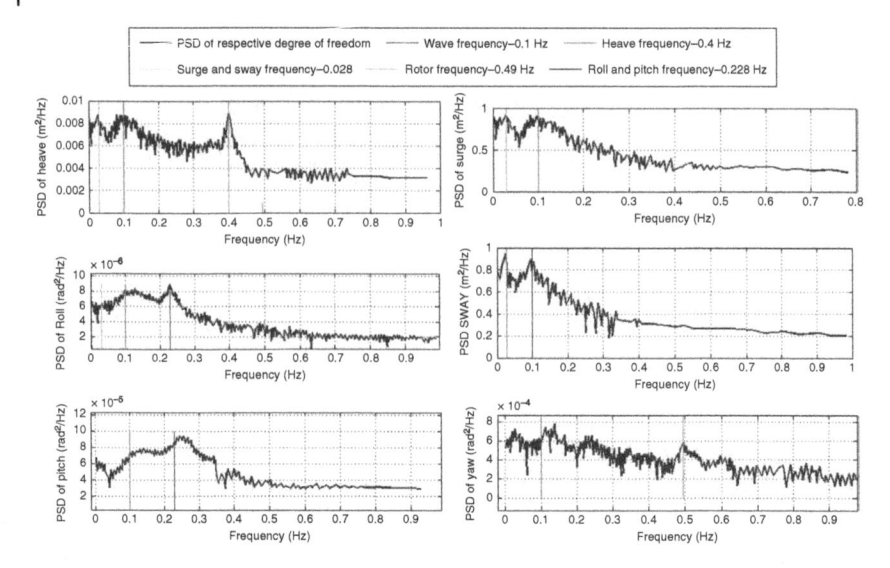

Figure 5.6 PSD plots for different DOF.

5.8.5 Tether Response and Service Life Estimation

Figure 5.7 shows the tension variation in the tethers of the three buoyant legs in the presence of waves corresponding to Douglas sea condition 3. The tether tension variation is not similar for the tethers of each buoyant leg due to the phase lag effect. The maximum tether tension variation lies within the range of 8.6–14.6%, given Douglas sea condition 3. Reduced tether tension variation confirms that the

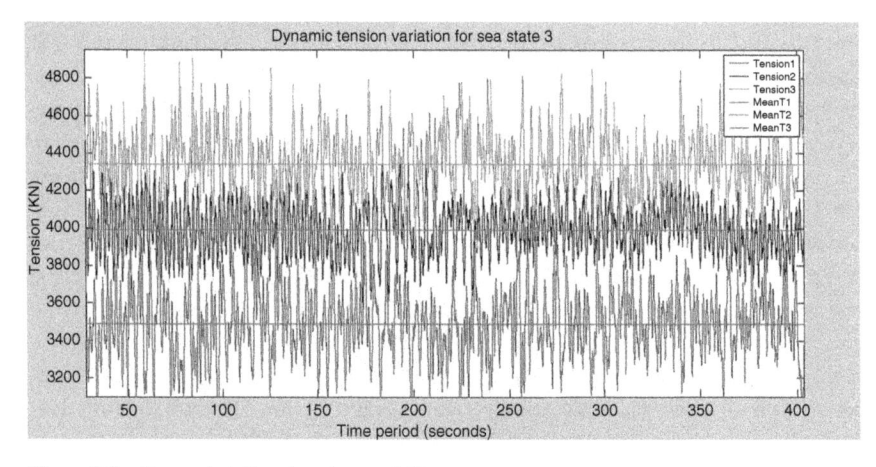

Figure 5.7 Dynamic tether tension variation.

Table 5.3 Service life estimation of the triceratops.

Sea conditions	Wind speed (knots)	Significant wave height (meters)	Fatigue life (in years)
3	11–16	0.5–1.25	27
6	28–33	4–6	9.47

probability of tether pullout is reduced. However, the tethers may underdo fatigue failure. Fatigue damage and service life are calculated according to standard regulations and methodology as explained in Chapter 3. The fatigue damage of the tethers in 900 seconds is 3.014×10^{-6}, and the damage is extrapolated to 0.106 for one year. The damage is equivalent to 1 in 9.47 years, which amounts to the service life of the structure in very rough conditions. The fatigue life under different sea conditions is given in Table 5.3.

5.9 Stiffened Triceratops

In a stiffened triceratops, the buoyant legs of the basic triceratops are modified with three cylindrical columns that are interconnected to the central moon pool by a finite number of stiffeners. The plan and elevation of the stiffened buoyant legs are shown in Figure 5.8. Ring stiffeners within the buoyant legs, and external stiffeners that interconnect the buoyant legs, are provided at equal intervals along the length of the buoyant legs. These fabricated legs, once interconnected, are assumed to behave like a monolithic unit and thereby reduce the effect of the encountered wave loads on the platform legs.

Similar to the basic conceptual model of the triceratops, each stiffened buoyant leg is connected to the rectangular deck by three ball joints. The rectangular deck is another innovation attempted by researchers. The shape of the deck is modified to provide more work space on the topside. Both experimental and numerical analyses were carried out on this special type of offshore triceratops, in order to predict the exact behavior of the structure and to understand the difference in responses in comparison with the basic triceratops conceptual model.

5.9.1 Preliminary Design

The preliminary design of the geometric form can be subsequently assessed using a hydrostatic analysis of the scale model. Scaling laws must be followed while carrying out experimental studies. This ensures that the ratios of unlike forces in the prototype and model remain similar. For example, Froude scaling is an appropriate

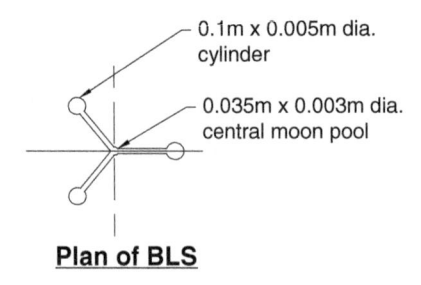

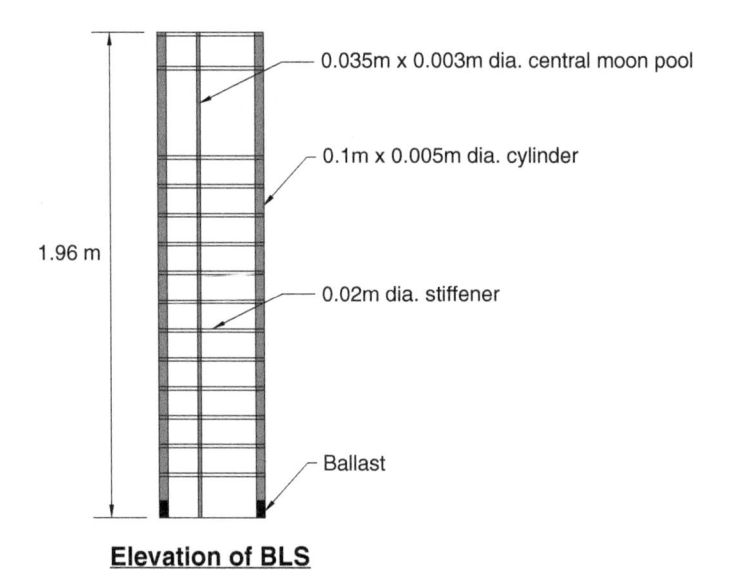

Figure 5.8 Plan and elevation of a stiffened buoyant leg.

scaling law for inertia and gravity forces and is commonly used for experimental investigations on scale models. Froude scaling confirms the similitude of inertia and gravity forces, but not viscous and pressure forces. The desired radius of gyration, mass distribution, and global stiffness must be scaled with respect to those of the prototype. The geometric parameters of the prototype and 1 : 75 scale experimental model are given in Table 5.4.

Based on the hydrostatic parameters, fabrication of the triceratops with interconnected buoyant legs must be carried out. The spacing of the transverse stiffeners and ring stiffeners plays an important role because it controls the mass ratio of the topside to the buoyant legs. Spacing and configuration of the stiffeners should be chosen to meet the desired design draft given free-floating conditions. Mass comparisons of individual components of the triceratops are presented in Table 5.5.

Table 5.4 Geometric parameters of the offshore triceratops.

Description	Experimental Model (m)	Prototype (m)
Buoyant leg		
Length of cylinder (3+1)	1.96	147
Outer diameter of cylinder (3)	0.1	7.5
Inner diameter of cylinder (3)	0.09	6.75
Thickness of cylinder	0.005	0.375
Outer diameter of central moon pool	0.035	2.625
Inner diameter of central moon pool	0.029	2.175
Thickness of moon pool	0.003	0.225
Ballast		
Length	0.08	6
Diameter	0.085	6.375
Tether		
Length	2.8	210
Diameter	0.001	0.075
Deck		
Length	0.91	68.25
Breadth	1.0	75.0
Thickness	0.014	1.05

The fabricated model, as per the specifications mentioned in Table 5.4, is shown in Figure 5.9. Buoyant leg units with stiffeners are fabricated using an acrylic material, while the ballast weights are made of lead. Wood is used in the fabrication of the rectangular deck structure, and stainless steel tethers are used to moor the buoyant legs to the seafloor. The ball joints shown in Figure 5.10 are fabricated using stainless steel.

5.9.2 Response to Wave Action

The results of the experimental investigations carried out on the scale model for different wave heights and wave periods are described next. The surge, heave, and pitch RAOs of the deck over the buoyant leg units for 0° incident waves are shown in Figure 5.11. The displacement of the structure in the surge and heave directions with respect to the increase in wave height is almost constant. As the wave period increases, the surge and heave responses of the deck structure increase

Table 5.5 Mass properties of the triceratops.

Description	Stiffened buoyant leg model (kg)	Prototype (ton)	Triceratops model (kg)	Prototype (ton)
Topside				
Drilling system	2.9	1223.44	–	–
Other systems	11.16	4708.13	–	–
Allowance	0.5	210.94	–	–
Steel	8.44	3560.63	–	–
Total mass of topside	23	9703.13	1.2	4059
Buoyant leg				
Steel	43.5	18 351.56	5.4	18 225
Ball joint and appurtenances	3.3	1392.19	0.3	1013
Ballast	45	18 984.38	6.23	21 036
Pretension + tether mass	17	7171.88	2.43	8653
Total mass of buoyant leg	108.8	45 900	14.36	48 465
Displacement	118.79[a]	50 114.53	15.56[a]	52 982

[a] Considering flume water density, $\rho = 1016 \text{ kg/m}^3$.

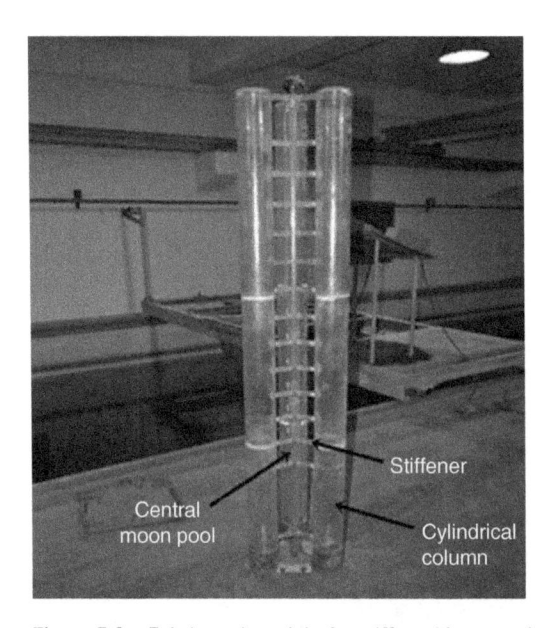

Figure 5.9 Fabricated model of a stiffened buoyant leg.

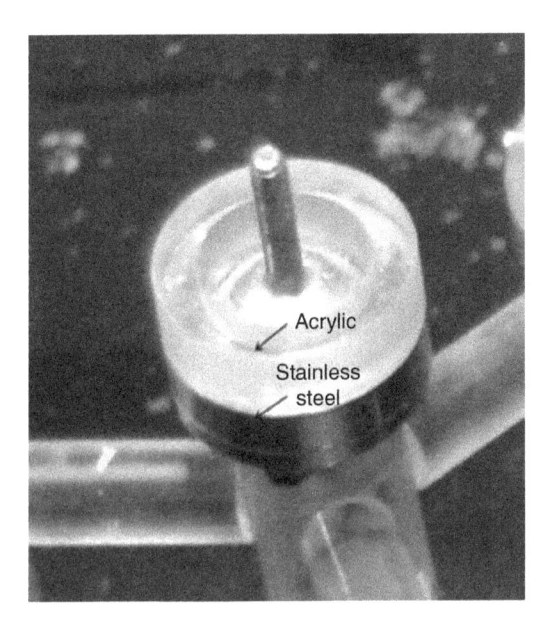

Figure 5.10 Fabricated model of a ball joint.

steadily. The difference in the surge RAO values of the deck and buoyant legs increases with an increased wave period. However, in the case of the heave response, the difference in the RAO values of the deck and buoyant legs decreases with an increased wave period. The RAO results of the translational response of the deck and buoyant legs of the stiffened triceratops show that these values are directly proportional to the linear displacement of the structure. However, in the case of the rotational response, the RAO values are directly proportional to the structure's rotational displacement. Hence, if there is an increase in the wave period, there is a decrease in the structure's pitch response. On the other hand, the difference in the pitch RAO values of the deck and buoyant legs steadily increases with an increased wave period. There is a negligible effect on the deck and buoyant legs from an increase in wave height at lower wave periods. However, there is less effect due to wave height at higher wave periods only in the heave response. The pitch response of the deck is reduced as the wave period increases, and this makes the structure more suitable in ultra-deep water.

5.9.3 Effect of Wave Direction

The surge, heave, and pitch RAOs of the deck and buoyant legs for 90° incident waves are shown in Figure 5.12. The pitch response of the deck is reduced as the wave period increases. This ensures that the deck is less responsive than the buoyant legs in ultra-deep water, except in the heave DOF. The RAO plots for the surge, heave,

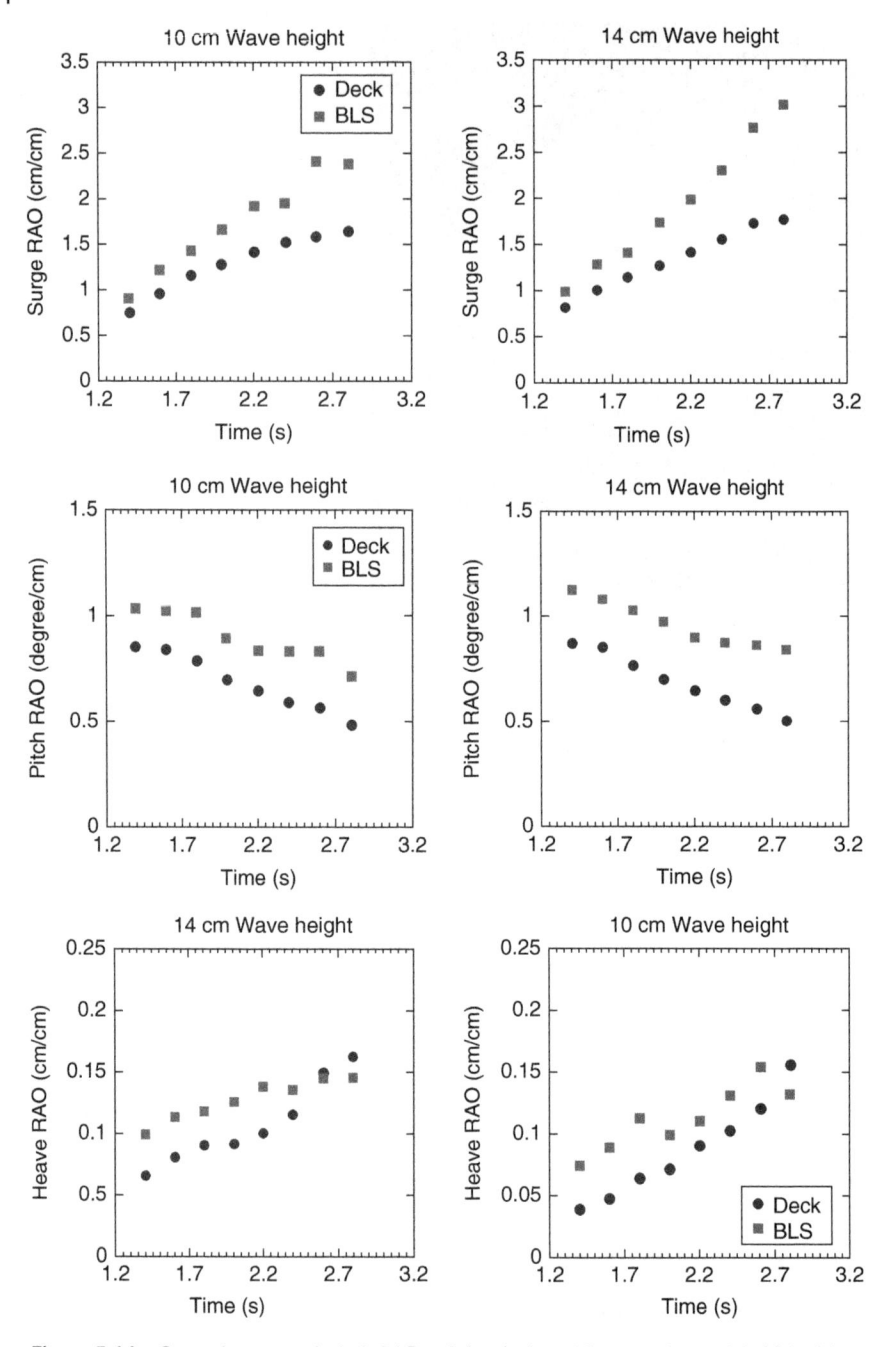

Figure 5.11 Surge, heave, and pitch RAOs of the deck and buoyant legs with 0° incident waves.

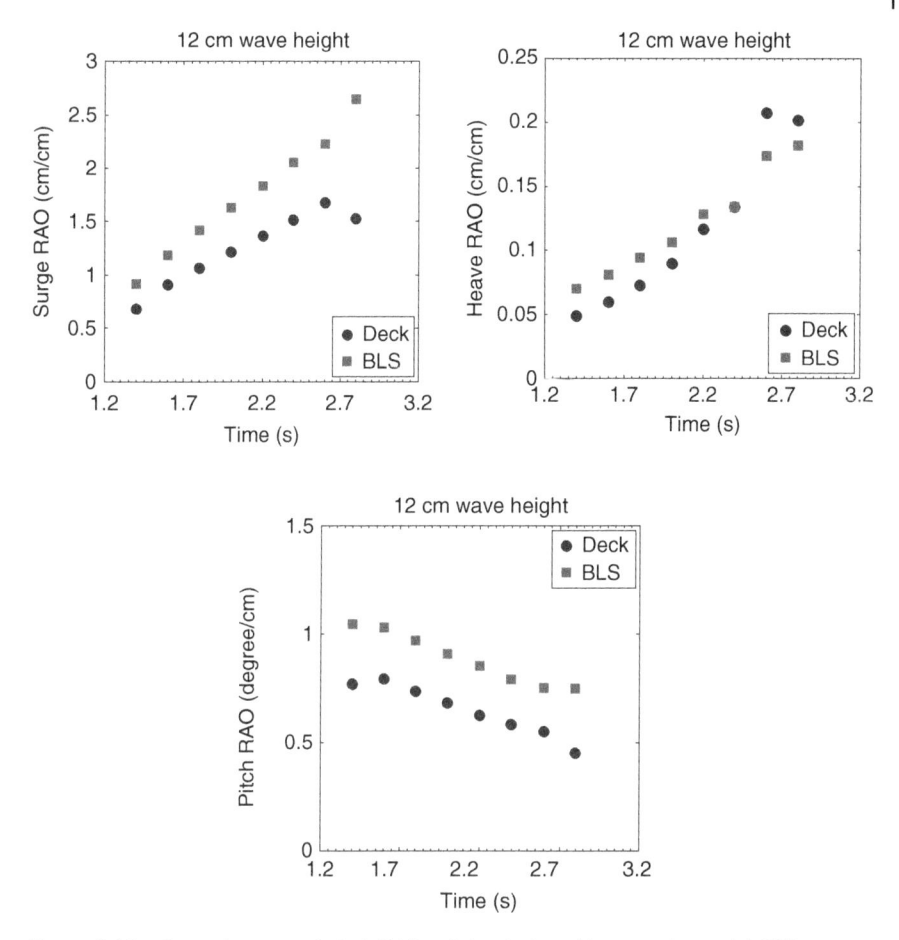

Figure 5.12 Surge, heave, and pitch RAOs of the deck and buoyant legs with 90° waves.

and pitch of the deck and buoyant legs are shown in Figure 5.13. The pattern of the graph is the same as that of the 0° and 90° wave incident angles. The change in wave direction shows minor variations in the surge and pitch responses of the deck independent of the buoyant leg response at higher wave periods. The effect of wave direction on the deck and buoyant legs of the stiffened triceratops is shown in Figure 5.14.

5.10 Triceratops with Elliptical Buoyant Legs

Large structures with elliptical cross sections are currently under consideration for use in offshore oil drilling and production platforms. In recent years, elliptical hollow steel sections have become popular due to their viability for

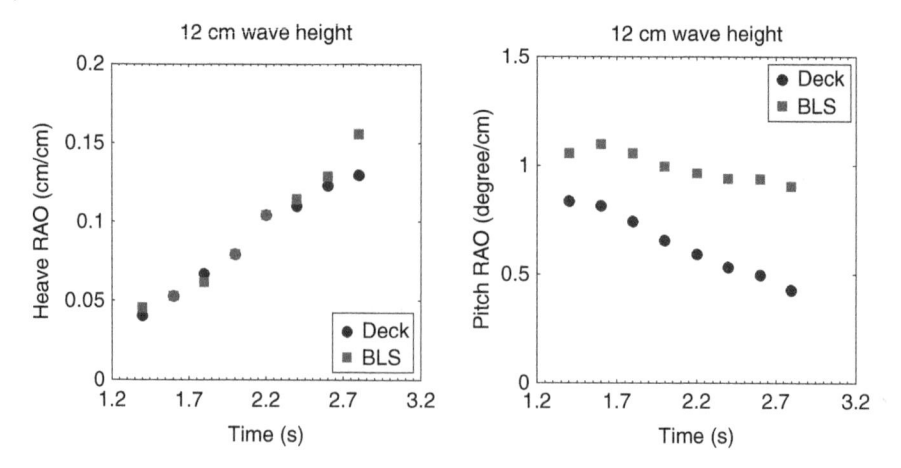

Figure 5.13 Surge, heave, and pitch RAOs of the deck and buoyant legs with 180° waves.

applications; the shapes are also included in the European Standard 10 210 (CEN-EN 2019; Silvestre 2008). The accurate determination of wave forces on elliptical structures is necessary for effective design. Several studies were reported on wave force calculations for elliptical cylinders using linear diffraction theory. The total wave force acting on an elliptical cylinder depends upon the wave incident angle and phase (Wang et al. 2019). In the case of an array of elliptical cylinders, the complex hydrodynamics affect the total response of the structure. The hydrodynamic interactions of the array of elliptical cylinders also induce a sway force even if the wave force action is parallel to the longitudinal axis of the ellipse (Chatjigeorgiou and Mavrakos 2010). With respect to buckling behavior, elliptical cylindrical shells with moderate to high eccentricity ($a/b \geq 2$) exhibit ultimate loads greater than their buckling loads because their post-buckling behavior is more stable than that of circular cylindrical shells (Booton et al. 1971).

5.10.1 Conceptual Development

The three buoyant legs of a triceratops usually have a circular cross section in order to overcome difficulties associated with fabrication. Because elliptical cross sections are being used in the offshore industry, the triceratops is conceptually modified with elliptical buoyant legs that have different eccentricities (a/b), as shown in Figure 5.15. The area of the elliptical cross section is the same as that of a circular buoyant leg with a 15.0 m diameter, in order to maintain the buoyancy of the triceratops. Based on the assumed eccentricities, the size of the elliptical cross section is calculated.

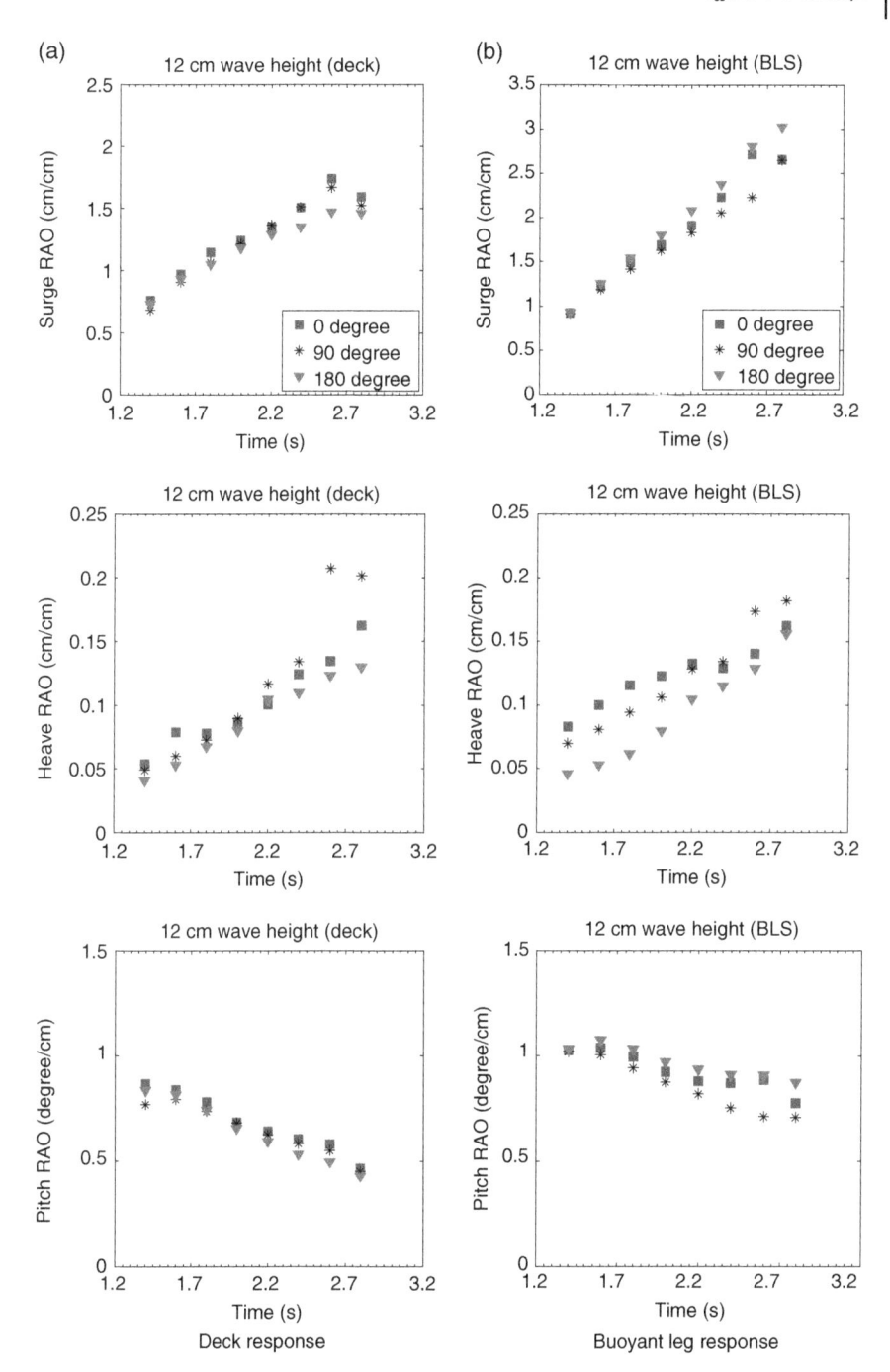

Figure 5.14 Effect of wave direction on the stiffened triceratops.

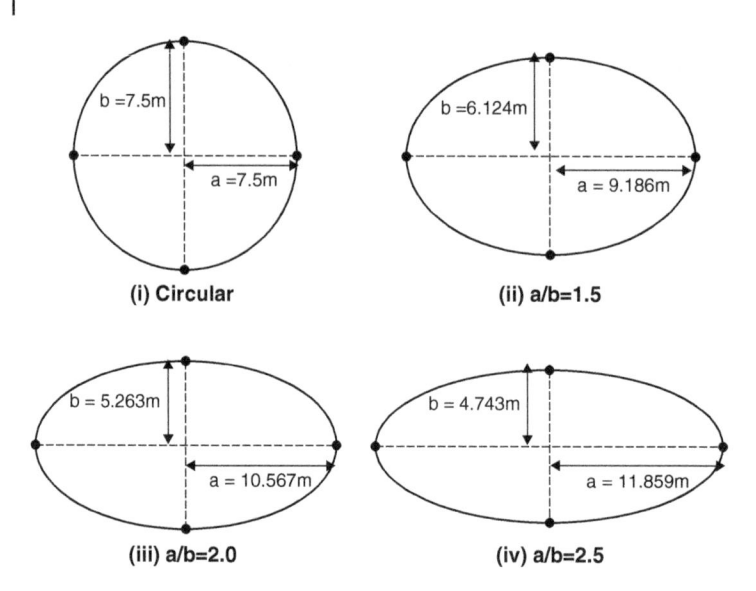

Figure 5.15 Cross section of the buoyant legs.

The numerical model of a triceratops is developed using ANSYS AQWA. The buoyant legs and the deck are modeled as separate entities with a corresponding center of gravity and point mass. The topside is modeled with three deck levels, using solid elements, and the buoyant legs are modeled using shell elements. Thus, the wave force acting on the buoyant legs is calculated according to diffraction theory. The deck and buoyant legs are connected using ball joints that restrict the transfer of rotational motion and only allow the transfer of translational motion between the deck and buoyant legs. The buoyant legs are connected to the seabed by taut-moored tethers, modeled as linear cables. The whole structure is then meshed using triangular and quadrilateral elements with program-controlled optimum meshing. A triceratops with elliptical buoyant legs is also numerically modeled as just explained. A plan view of the triceratops with elliptical buoyant legs with different eccentricities is shown in Figure 5.16.

5.10.2 Response of a Triceratops with Elliptical Buoyant Legs to Wave Action

A time-domain analysis is then carried out in the presence of the action of regular waves in rough sea conditions with wave height 2.0 m and wave period 5.0 seconds, in order to simulate the real-time motion of the triceratops. The positions and velocities of the deck and buoyant legs are obtained at each time step by integrating the accelerations developed by the environmental loads. The numerical

(i) Circular

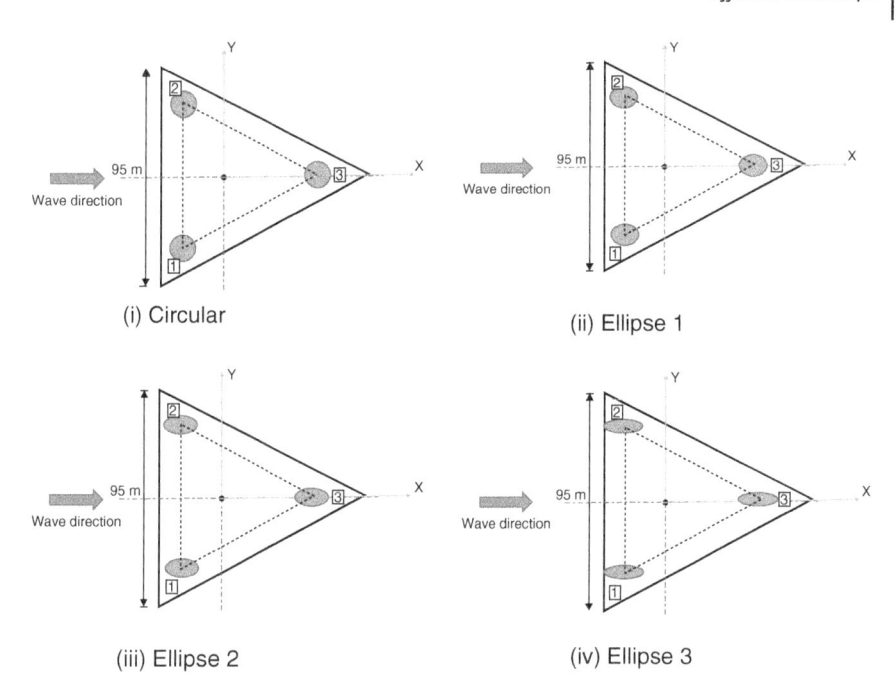

(ii) Ellipse 1

(iii) Ellipse 2

(iv) Ellipse 3

Figure 5.16 Plan view of the triceratops with circular and elliptical buoyant legs.

analyses of the triceratops with different buoyant legs are carried out at a 0° incident wave angle, as shown in Figure 5.16. The total responses of the deck in different DOF at a 0° incident wave angle are given in Table 5.6. In rough sea conditions, the total surge response of the deck for ellipse 1 increases by 26% compared to that of the triceratops with circular buoyant legs. The increase in the deck response of the triceratops with elliptical buoyant legs in comparison with the triceratops with circular buoyant legs is reduced with increased eccentricity of the buoyant legs. With circular buoyant legs, the deck sway response is about 4.70% of the surge response. With elliptical buoyant legs, the sway response is less when compared to the surge response only for ellipse 2. The transverse vibration in the buoyant legs is less for ellipse 2, with an a/b ratio of 2.0. The transverse vibration is found to be high for ellipse 3, where the sway response is about 37.80% of the surge response. The heave, roll, pitch, and yaw responses of the deck decrease with increased eccentricity of the buoyant legs.

The response of circular and elliptical buoyant legs and their complex interactions with the deck through ball joints affects the total response of the platform. The surge response of buoyant leg 1 in all of the ellipse cases is greater than in the circular case. The buoyant leg response for ellipse 3 is less than in the other ellipse cases but 2.5% greater than in the circular case. Similar to the deck response, the

Table 5.6 Response of the triceratops given rough sea conditions.

Component	Degree of freedom	Circular	Ellipse 1	Ellipse 2	Ellipse 3
Deck	Surge (m)	0.4047	0.5098	0.4296	0.4067
	Sway (m)	0.0192	0.1434	0.0348	0.1537
	Heave (m)	0.0012	0.0012	0.0011	0.0011
	Roll (deg)	0.0035	0.0031	0.0027	0.0028
	Pitch (deg)	0.0037	0.0038	0.0036	0.0034
	Yaw (deg)	0.0253	0.3688	0.0815	0.0293
Buoyant leg 1	Surge (m)	0.2910	0.3660	0.3438	0.2983
	Sway (m)	0.0284	0.2885	0.0805	0.1575
	Heave (m)	0.0028	0.0027	0.0029	0.0027
	Roll (deg)	0.0056	0.0800	0.0050	0.0086
	Pitch (deg)	0.1290	0.1272	0.0883	0.0969
	Yaw (deg)	3.8035	1.8000	0.9267	0.8999

transverse vibration of the buoyant leg is found to be less for ellipse 2 with eccentricity 2.0. The sway response in this case is only about 23% of the surge response. Comparatively less yaw response is observed in the elliptical buoyant legs than the circular legs: the yaw response for ellipse 2 is only 24% of the yaw response in the circular case. In high sea conditions, the surge response is less for ellipse 2, similar to the deck response. This shows the rigidity of the platform in the horizontal plane, where translation motions are transferred by the ball joints from the buoyant legs to the deck.

The total force developed on the buoyant legs is highly affected by the shape of the buoyant legs. The maximum total force along the x-axis is observed for ellipse 3. Comparing the elliptical buoyant legs, the force along the y-axis is less for ellipse 2. The force in the horizontal plane is at a maximum in the elliptical cases, whereas the force in the vertical plane is at a maximum in the circular case. Due to reduced drag in the circular configuration, the total moment about the x-axis is less in comparison with the elliptical buoyant legs. However, the total moment about the y-axis is at a maximum in the circular case. Comparatively less moment about the y-axis is observed for ellipse 3. Increased moment about the z-axis is observed in the circular case. The total force–time history in the surge, heave, and pitch DOF in rough sea conditions is shown in Figure 5.17.

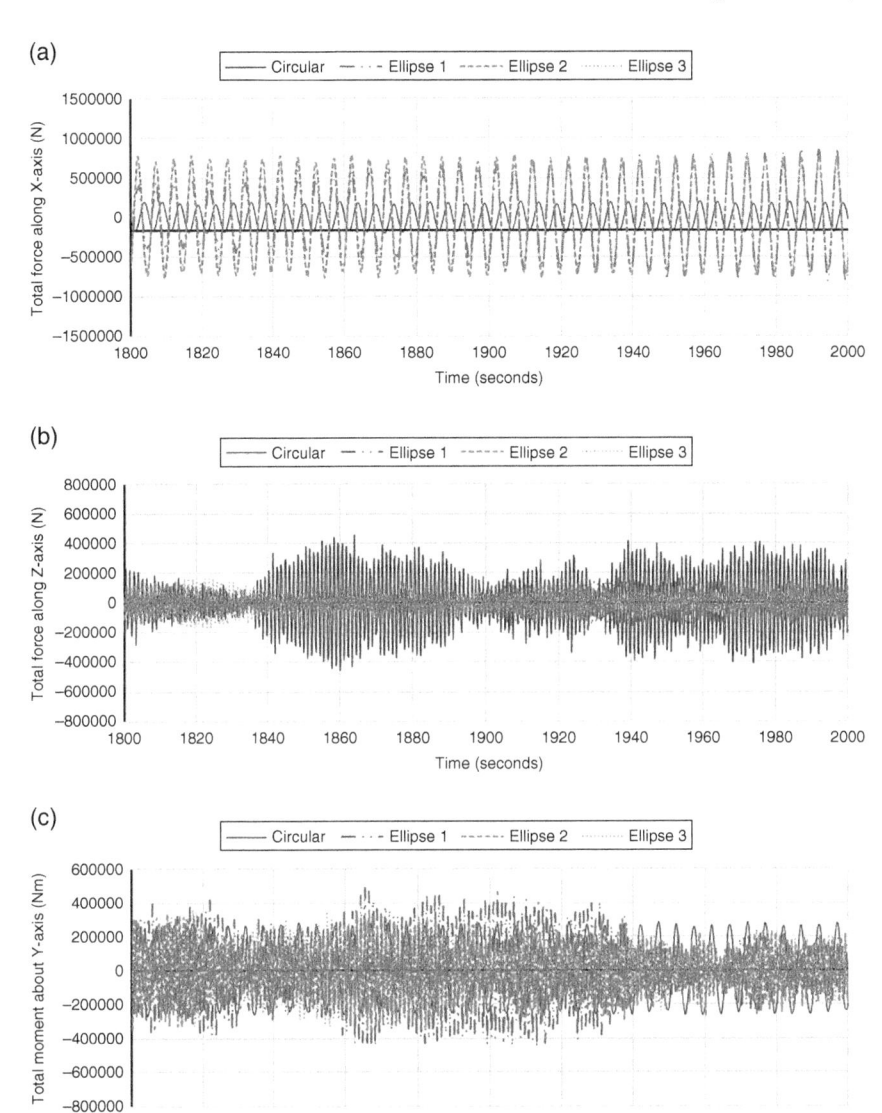

Figure 5.17 Total force–time history, given high sea conditions.

5.11 Summary

In a triceratops-based wind turbine, the environmental loads fail to excite the system in any DOF, which reinforces the dynamic stability of the system. The frequency responses of the triceratops in all flexible DOF are similar qualitatively, while there is a marked difference in the roll and pitch responses. The service lives of tethers in operational and very rough sea conditions are 27 years and 9.47 years, respectively. In the stiffened triceratops, the surge, heave, and pitch responses of the buoyant legs are reduced with increased pre-tension in the tethers. The deck response is further reduced with the help of ball-joint connections between the deck and buoyant legs. In addition, the surge and pitch responses of the buoyant legs and deck of the triceratops model are not influenced by the wave direction, and the variations in the responses given different wave directions are acceptably marginal. The heave RAO of the deck with a 90° wave direction is higher compared to the responses with 0° and 180° wave directions, due to the effect of higher set-down of the structure at higher wave periods, which occurs at deepwater and ultra-deepwater depths. Elliptical buoyant legs with an eccentricity greater than 2.0 are found to be advantageous in alleviating wave loads effectively. An offshore triceratops therefore derives an advantage from the chosen geometric form in addition to being advantageous for deepwater and ultra-deepwater applications. Hence, the platform geometry is seen as an effective alternative for ultra-deep water.

Model Exam Paper 1

Maximum marks: 100   Time: 3 hours
Answer all questions.
No additional support materials are permitted.

PART A (20 × 1 = 20)

1. _____ are good to transfer applied loads to the foundations of the platform.
2. Compliant platforms are based on a _____ concept of design.
3. The design of the geometric form of compliant structures is principally dominated by balancing the _____ and _____ of the platform.
4. The _____ is a unique component of an articulated tower, which connects it to the foundation system.
5. _____ use the friction between two surfaces to dissipate energy.
6. The natural frequency of a system can be measured experimentally using _____.
7. The yaw response of the deck is attributed to the _____ in the recentering capability of the buoyant legs under directional wave loads.
8. Due to fluctuating winds, compliant offshore structures are more susceptible to _____.
9. The steady wind velocity is measured at _____.
10. Wave and wind spectra are _____ and _____ banded, respectively.
11. Because a triceratops is a _____ structure, tether failure will not result in the complete collapse of the structure.

Offshore Compliant Platforms: Analysis, Design, and Experimental Studies,
First Edition. Srinivasan Chandrasekaran and R. Nagavinothini.
© 2020 John Wiley & Sons Ltd.
This Work is a co-publication between John Wiley & Sons Ltd and ASME Press.

12. Large, continuous ice sheets called _____ are formed due to ice cakes freezing together.
13. The ice sheet that remains attached to the shore developed during winter is called _____.
14. Level ice action induces _____ in offshore structures.
15. The design guidelines suggest _____ as the minimum collision energy for the design of offshore structures subject to ship–platform collisions.
16. _____ is calculated by multiplying the effective ice pressure and the contact area.
17. _____ is an appropriate scaling law for inertia and gravity forces.
18. Stiffening the buoyant legs of a triceratops _____ the tether tension.
19. Cyclic loading in a structural system may lead to _____.
20. In a BLSRP, the buoyant legs are connected to the deck by _____.

PART B (10 × 3 = 30)

1. What are the factors that influence the choice of a geometric form for an offshore structure?
2. Explain TLP mechanics.
3. What are viscous fluid dampers?
4. Explain the structural action of a BLSRP.
5. How do you find the damping ratio using the logarithmic decrement method?
6. Why is yaw motion observed in the deck of a BLSRP with a 0° wave heading angle?
7. List the widely used wave spectra in the hydrodynamic analysis of structures.
8. Draw the wind-generated current velocity profile.
9. What are the effects of wind, waves, and currents on sea ice?
10. List the factors that affect the ice load acting on an offshore structure.

PART C (5 × 10 = 50)

1. Briefly explain the structural action of a tension leg platform (TLP).
2. Explain passive control devices.
3. Write briefly about current.
4. Explain the major subsystems in a wind turbine.
5. Expand the following: *FSO, FPSO, FPU, TLP, TLD, TLCD, TMD, MLAT, DVA, BLSRP*.

Model Exam Paper 1: KEY

Maximum marks: 100 Time: 3 hours
Answer all questions.
No additional support materials are permitted.

PART A (20 × 1 = 20)

1. *Fixed structures* are good to transfer the applied loads to the foundations.
2. Compliant platforms are based on a *relative displacement* concept of design.
3. The design of the geometric form of compliant structures is principally dominated by balancing the *buoyancy force* and *weight* of the platform.
4. The *universal joint* is a unique component of an articulated tower, which connects it to the foundation system.
5. *Frictional dampers* use the friction between two surfaces to dissipate energy.
6. The natural frequency of a system can be measured experimentally using *free-decay tests.*
7. The yaw response of the deck is attributed to the *time delay* in the recentering capability of the buoyant legs under directional wave loads.
8. Due to fluctuating winds, compliant offshore structures are more susceptible to *low-frequency oscillations*.
9. The steady wind velocity is measured at *10.0 m above MSL*.
10. Wave and wind spectra are *narrow* and *wide* banded, respectively.
11. Because a triceratops is a *positively buoyant* structure, tether failure will not result in the complete collapse of the structure.
12. Large, continuous ice sheets called *ice floes* are formed due to ice cakes freezing together.
13. The ice sheet that remains attached to the shore developed during winter is called *shore-fast ice*.
14. Level ice action induces *random vibrations* in offshore structures.
15. The design guidelines suggest *4.0 MJ* as the minimum collision energy for the design of offshore structures subject to ship–platform collisions.
16. *The maximum crushing ice force* is calculated by multiplying the effective ice pressure and the contact area.
17. *Froude scaling* is an appropriate scaling law for inertia and gravity forces.
18. Stiffening the buoyant legs of a triceratops *increases* the tether tension.
19. Cyclic loading in a structural system may lead to *fatigue failure.*
20. In a BLSRP, the buoyant legs are connected to the deck by *hinged joints*.

PART B (10 × 3 = 30)

1. What are the factors that influence the choice of a geometric form for an offshore structure?
   a) Structural form with a stable configuration
   b) Geometric form leading to low installation, fabrication, and decommissioning costs
   c) Geometric form that requires a lower CAPEX and leads to a high return on investment (ROI)
   d) Geometric form that can result in an early start for production and that possesses high mobility
   e) Geometric form that requires the least possible intervention, so that uninterrupted production can take place

2. Explain TLP mechanics.
   When no load is applied on the structure, the structure is in a stationary, stable condition. Due to excess buoyancy, the tethers are in high tension so that the platform is held down to the seabed. Under lateral loading from wind, waves, or currents, the structure experiences a lateral displacement. The lateral displacement of the TLP is called offset. The offset condition of the TLP pulls down the structure. The vertical displacement is called setdown.

3. What are viscous fluid dampers?
   - A viscous fluid damper is similar to a conventional shock-absorber. It consists of a closed cylinder-piston, which is filled with fluid (usually silicon oil).
   - These dampers are typically installed as diagonal braces in building frames (preferably steel structures).
   - To provide optimal damping, buildings are often equipped with multiple dampers in place of diagonal beams on every floor.

4. Explain the structural action of a BLSRP.
   The structural action of a BLSRP under lateral loads is similar to other compliant offshore structures:
   a) It is similar to a TLP, because a restraining system with tethers is common.
   b) It is similar to a spar platform, because each buoyant leg resembles a spar buoy due to the deep draft.
   c) It has articulated towers due to the presence of hinged joints. Therefore, a BLSRP is a hybrid compliant platform.

5. How do you find the damping ratio using the logarithmic decrement method?

The damping ratio in the respective DOF is obtained by the logarithmic decrement method and is as follows:

$$\delta = \frac{1}{n} \ln\left(\frac{x_0}{x_n}\right)$$

where x_0 is the higher value of the two peaks. x_n is the value of the peak after n cycles. The damping ratio is determined by:

$$\zeta = \frac{1}{\sqrt{1 + \left(\frac{2\pi}{\delta}\right)^2}}$$

6. Why is yaw motion observed in the deck of a BLSRP with a 0° wave heading angle?

   As buoyant legs are symmetrically spread with respect to the vertical axis of the platform, it is imperative to envisage a non-uniform phase lag in the recentering process; this causes the yaw motion of the deck.

7. List the widely used wave spectra in the hydrodynamic analysis of structures.

   a) Pierson-Moskowitz (PM)
   b) JONSWAP
   c) ISSC
   d) Bredneidger
   e) Ochi-Hubble

8. Draw the wind-generated current velocity profile.

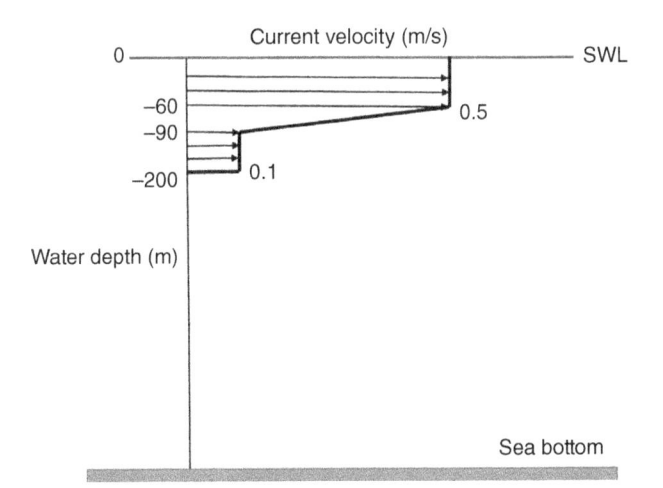

9. What are the effects of wind, waves, and currents on sea ice?
   - *The exposure of developed ice to the wind, waves, and currents lead to the deformation of the ice along with an increase in the brittleness of the ice crust.*
   - *The continuous action of wind, waves, and currents transforms this continuous flat ice sheet into pressure ice fields with rough surfaces.*

10. List the factors that affect the ice load acting on an offshore structure.
    a) *Structural geometry of the platform*
    b) *Location and environmental conditions*
    c) *Ice properties such as thickness, velocity, and crushing strength*
    d) *Ice–structure interaction phenomena*

PART C (5 × 10 = 50)

1. Briefly explain the structural action of a tension leg platform (TLP).

   Tension leg platforms (TLPs) operate successfully in deep water. They are classified as hybrid compliant structures that are suitable for both drilling and production operations. Compliancy of the platform is restricted to the surge, sway, and yaw DOF, while the heave, roll, and pitch DOF remain stiff. This compliancy is obtained using taut-moored pre-tensioned tethers that hold the platform in position. The tethers connect the platform column to the seabed through the pile templates. The axial stiffness of the tether is kept significantly high to counteract the excessive buoyancy inherent in the design. Tension in the tethers minimizes the motion of the platform in the vertical plane, while excessive buoyancy helps the tethers to remain stiff; hence the name TLP. The TLP concept is designed in such a manner that the natural periods of the platform are either too low or too high in comparison to the operational wave periods.

2. Explain passive control devices.

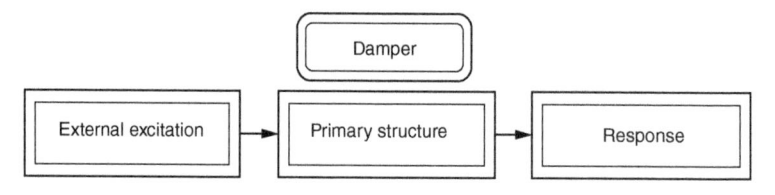

   ```
   External excitation → Primary structure → Response
                              ↑
                           Damper
   ```

   Passive systems require no external energy for successful operation, which is one of the major advantages of such systems in comparison to other types. A key benefit of passive control devices is that once installed in a structure, they do not require any startup or operation energy, unlike active and semi-active systems. Passive control devices are active at all times until maintenance, replacement, or dismantling is required.

Passive control systems include friction dampers, metallic yield dampers, and viscous fluid dampers. Alternative types of passive control systems contain a spring (or spring-like component), which is tuned to a particular natural frequency of the structure for maximum damping. Examples of these passive control devices are tuned mass dampers, tuned liquid dampers, and tuned liquid column dampers.

3. Write briefly about current.
   Current generation in the sea is mainly due to the following factors:
   - *Wind effects*
   - *Tidal motion*
   - *Temperature differences*
   - *Density gradients*
   - *Salinity variations*

   The apparent wave period and the total water particle velocity are altered by the presence of currents. The current action also imposes additional drag forces on structures, which in turn affects the tether tension variation of compliant structures. Wind-generated currents are highly concentrated close to the sea surface, and the effect decreases with increased water depth. The current effect is included in the analysis by representing the current velocity, which varies linearly from the maximum value at the sea surface to zero at the seabed. The maximum current velocity of wind-generated current can be approximated as 1.0–3.0% of the sustained wind velocity. The current in the same direction as the waves increases the wavelength and the wave period. The increased wave period (10%) due to current action is called the apparent wave period.

4. Explain the major subsystems in a wind turbine.
   The seven major subsystems in a wind turbine are as follows:
   - *Blades*
   - *Nacelle*
   - *Controller*
   - *Generator*
   - *Rotor*
   - *Tower*
   - *Floating body*

   The rotor houses a number of blades that determine the system performance of the wind turbine. A three-bladed upwind design is predominantly used in the design of the rotor, and the blade design is usually based on the pitch control. The nacelle protects the generator, controller, gearbox, and shafts. The tower supports the wind turbine nacelle and rotor. The total height of the tower at the particular site is usually governed by the rotor diameter and the nature of the loading conditions. Generators are used to convert the raw mechanical work of the wind turbine to useful electrical output. Changes in the blade pitch angle, generator loading, and nacelle yaw are monitored by the control system. The generated electrical output is transferred to a suitable electrical grid through cables buried in the seabed. If this method is uneconomical, in recent years the generated power has been transferred through battery storage. The wind turbine elements are supported on a floating body, and mooring systems are usually employed to position restrain the system.

5. Expand the following:
FSO – Floating storage and offloading
FPSO – Floating production, storage, and offloading
FPU – Floating production unit
TLP – Tension leg platform
TLD – Tuned liquid damper
TLCD – Tuned liquid column damper
TMD – Tuned mass damper
MLAT – Multi-leg articulated tower
DVA – Dynamic vibration absorber
BLSRP – Buoyant leg storage regasification platform

Model Exam Paper 2

Maximum marks: 100   Time: 3 hours
Answer all questions.
No additional support materials are permitted.

PART A (20 × 1 = 20)

1. _____ are insensitive under lateral loads arising from wind, waves, and currents.
2. Why is recentering considered important in the design of compliant structures?
3. Articulated towers are similar to TLPs, with tethers replaced by a _____.
4. Metallic yield dampers are known to have _____ behavior.
5. When the frequency of the tank motion is closer to one of the natural frequencies of the tank fluid, _____ occurs.
6. Higher stiffness in the yaw motion of a BLSRP is due to the _____ of the buoyant legs.
7. Because the BLSRP is positive-buoyant, _____ on the tethers is necessary to ensure position restraint.
8. The wind load acting on the deck of the platform will induce additional moment, resulting in _____.
9. The maximum current velocity of the wind-generated current can be approximated as _____ of the sustained wind velocity.
10. With an increase in the severity of the sea conditions, the _____ in the surge DOF of the triceratops increases.
11. The design of offshore structures is mainly governed by these forms of ice: _____, _____, and _____.
12. When the ice force frequency becomes equal to the natural frequency of the structure, the _____ of the structural force will be high.
13. In the case of a three-legged structure like a triceratops, the maximum ice force occurs when ice acts on _____.
14. In a wind turbine, the _____ houses a number of blades that determines the system performance.
15. Current action imposes additional _____ forces on the structure.
16. A _____ is a passive type of damper that imposes response control using the principal of inertia.
17. The heave natural period is reduced by _____ the pipe wall thickness of the tethers.

18. Among the wind spectra, lower spectral energy is observed in the _____.
19. Compliant platforms are position-restrained by _____.
20. Hinged joints also serve as _____, which controls the deck motion even for a large movement/rotation of the buoyant legs.

PART B (10 × 3 = 30)

1. How do you classify offshore structures based on station-keeping characteristics?
2. Name some semi-active control devices.
3. What are the advantages of a MLAT?
4. What is the major reason for the differences in responses of buoyant legs?
5. Write the canonical form of the Mathieu equation.
6. List the wind spectra used to represent random wind for the analysis of structures.
7. Differentiate pancake ice and ice cakes.
8. Explain the formation of icebergs.
9. Differentiate limit stress and limit force failure.
10. List the disadvantages of offshore wind turbines.

PART C (5 × 10 = 50)

1. Write briefly about the spar platform.
2. Describe tuned mass dampers.
3. How is the service life of the structure calculated based on fatigue analysis?
4. How does an increase in temperature affect material properties?
5. Expand the following: *FSRU, FLNG, PM, IFFT, API, RAO, IEC, HAWT, VAWT, RNA*

Model Exam Paper 2: KEY

Maximum marks: 100 **Time: 3 hours**
Answer all questions.
No additional support materials are permitted.

PART A (20 × 1 = 20)

1. *Fixed platforms* are insensitive under lateral loads arising from wind, waves, and currents.

2. Why is recentering considered important in the design of compliant structures?
   It is very important in the context of compliant platform design because large displacements are essentially permitted as a part of the design itself.
3. Articulated towers are similar to TLPs, with tethers replaced by a *single high-buoyancy shell*.
4. Metallic yield dampers are known to have *stable hysteretic* behavior.
5. When the frequency of the tank motion is close to one of the natural frequencies of the tank fluid, *large-amplitudes sloshing* occurs.
6. Higher stiffness in the yaw motion of a BLSRP is due to the *symmetric layout* of the buoyant legs.
7. Because the BLSRP is positive-buoyant, *high initial pre-tension* on the tethers is necessary to ensure position restraint.
8. The wind load acting on the deck of the platform will induce additional moment, resulting in *an excessive pitch response*.
9. The maximum current velocity of the wind-generated current can be approximated as *1.0–3.0%* of the sustained wind velocity.
10. With the increase in the severity of the sea conditions, the *mean shift* in the surge DOF of the triceratops increases.
11. The design of offshore structures is mainly governed by these forms of ice: *ice sheets*, *pack ice*, and *icebergs*.
12. When the ice force frequency becomes equal to the natural frequency of the structure, the *dynamic amplification* of the structural force will be high.
13. In the case of a three-legged structure like a triceratops, the maximum ice force occurs when ice acts on *two buoyant legs simultaneously*.
14. In a wind turbine, the *rotor* houses a number of blades that determine the system's performance.
15. Current action imposes additional *drag* forces on the structure.
16. A *tuned mass damper (TMD)* is a passive type of damper that imposes response control using the principle of inertia.
17. The heave natural period is reduced by *increasing* the pipe wall thickness of the tethers.
18. Among the wind spectra, lower spectral energy is observed in the *Davenport spectrum*.
19. Compliant platforms are position-restrained by *tethers.*
20. Hinged joints also serve as *isolators*, which controls the deck motion even for a large movement/rotation of the buoyant legs.

PART B (10 × 3 = 30)

1. How do you classify offshore structures based on station-keeping characteristics?
   Fixed, compliant, and floating types

2. Name some semi-active control devices.
   Variable orifice fluid dampers, controllable friction devices, variable stiffness devices, controllable fluid dampers, and magneto-rheological dampers

3. What are the advantages of a MLAT?
   The payload and deck areas can be increased and made comparable to conventional production platforms in moderate water depths, and the sway or horizontal displacement of the deck is considerably reduced compared to single-leg articulated towers.

4. What is the major reason for the differences in responses of buoyant legs?
   Differences in the responses of buoyant legs are due to the variable submergence effect, which is one of the primary sources of nonlinearity in the excitation force.

5. Write the canonical form of the Mathieu equation.
   The Mathieu equation is a special form of the Hill equation, with only one harmonic mode. The canonical form is given by:

$$\frac{d^2 f}{d\tau^2} + \left(\delta - q\, Cos(2\tau)\right) f = 0$$

   where δ and q are Mathieu parameters, which are problem-specific.

6. List the wind spectra used to represent random wind for the analysis of structures.
   - Davenport spectrum
   - Harris spectrum
   - Kaimal spectrum
   - Simiu spectrum
   - Kareem spectrum
   - American Petroleum Institute (API) spectrum

7. Differentiate pancake ice and ice cakes.
   Circular ice pieces of diameter up to 3.0 m are called pancake ice, and larger pieces are called ice cakes. Pancake ice causes impact forces on offshore structures, which increase with increased wave height and current field.

8. Explain the formation of icebergs.
   Icebergs form due to the flow of glaciers followed by chunks of ice breaking due to the buoyancy of water. The direction and amplitude of wind and

currents govern the velocity of the icebergs in a particular location. The temperature variation above and below the water surface causes non-uniform melting of icebergs, which results in icebergs tilting, capsizing, and breaking. Breakage of icebergs leads to the formation of smaller bergs called growlers or bergy bits.

9. Differentiate limit stress and limit force failure.
   - In the case of limit stress failure, ice sheet failure occurs at the ice–structure interface when the environmental forces acting on the ice are greater than the failure strength of the ice. The common modes of ice failure given limit stress failure conditions are buckling and crushing.
   - In the case of limit force failure, the ice failure occurs far from the ice–structure interface, and the environmental forces acting on the ice sheet lead to the formation of ice ridges.

10. List the disadvantages of offshore wind turbines.
    - Very high initial investment
    - Complications involving the construction of the foundation and supporting structure, commissioning, and decommissioning
    - Less accessibility compared to onshore wind farms, which in turn increases downtime and increases the cost of maintenance and operation
    - Complexities arising due to the extreme hydrodynamic and aerodynamic loads acting on the supporting structures and turbines

PART C (5 × 10 = 50)

1. Write briefly about the spar platform.
   A spar platform is a large, deep-draft, cylindrical floating caisson, generally used for exploration and production purposes and installed at water depths of a few thousand meters. A spar has a long cylindrical shell called a hard tank, which is located near the water level. It generates high buoyancy for the structure, which helps keep the platform stable; the midsection is annulled and free flooding. The bottom part is called a soft tank and is utilized for placing the fixed ballast. It essentially floats the structure during transport and installation. In order to reduce the weight, drag, and cost of the structure, the midsection is designed to be a truss structure. To reduce the heave response, horizontal plates are introduced between the truss bays. The cell spar is the third generation of spar platforms, which was commissioned in 2004. It has a number of ring-stiffened tubes that are connected by horizontal and vertical plates. The hull is transported to the offshore site horizontally on its side.

2. Describe tuned mass dampers.

A tuned mass damper (TMD) is a passive type of damper that imposes response control using the principle of inertia. A TMD applies indirect damping to the structural system. The inertial force of the damper is equal to and opposite the excitation force for optimum control. TMDs are used for structures under lateral loads. A TMD consists of a secondary mass attached to the main structure through a spring-dashpot arrangement.

The energy of the primary structure is dissipated by inertial forces produced by the damper. The damper produces an inertial force in the direction opposite that of the structure's motion. The inertial force in the opposite direction helps reduce the motion of the primary structure. For maximum response reduction, the parameters of the TMD need to be tuned with those of the primary structure. The support system for the mass and tuning the frequency are important issues in the design of TMDs. While the mass of the damper is taken as a small fraction of the total mass of the primary structure (usually 1–5%), one of the main limitations is its sensitivity to the narrow frequency band of control. Mistuning of the TMD reduces its effectiveness considerably.

3. How is the service life of the structure calculated based on fatigue analysis?

The steps involved in the fatigue analysis of tethers are as follows:

Step 1: Dynamic response analysis of the triceratops
The dynamic response analysis of the triceratops should be carried out under the action of either environmental loads or accidental loads through experimental or numerical investigations.

Step 2: Tether tension variation
The tension variation of a tether should be obtained from the investigations carried out on the structure.

Step 3: Tether stress time history
From the known area of the tether and the tether tension variation, the tether stress variation time history is obtained.

Step 4: Stress histogram
The stress histogram should be developed from the stress time history. This stress histogram gives the stress range with the number of cycles.

Step 5: Allowable stress cycles
The allowable stress cycles should be calculated according to the standard regulations using the S-N curve approach. It is given by:

$\log N = \log B - m \log S$

where N is the number of allowable cycles, S is the stress range, and B and m are constants obtained from the S-N curves.

Step 6: Fatigue damage assessment

The fatigue damage of the tether is then calculated using the Miner-Palmgren rule given by:

$$D_f = \sum_{i=1}^{m} \frac{n_i}{N_i}$$

where Df is the fatigue damage, n is the number of stress counts from the histogram, and N is the number of allowable cycles from the S-N relationship.

Step 7: Service life calculation

Fatigue damage is then calculated for one year. Finally, the service life of the tethers is calculated by extrapolating the fatigue damage to one tether.

4. How does an increase in temperature affect material properties?

Offshore structures, especially the topsides, are constructed with different forms of steel. With respect to the grade of steel, the stress–strain characteristics vary significantly at elevated temperatures. Increased temperatures lead to thermal strains in the material, even in the absence of mechanical loading. So, the structural elements experience thermal strain without an increase in internal stresses under higher temperatures.

With the increase in temperature, Young's modulus, stiffness, and the yield strength of the structural steel decrease, with or without the development of mechanical strains. On the other hand, the material ductility increases, showing an indication of strength development.

In the case of mild carbon steel, the effective yield strength is reduced at higher temperatures (greater than 400 oC) at 2% strain, whereas the proportional limit and modulus of elasticity decrease with temperatures over 100 oC.

5. Expand the following:

FSRU – Floating storage and regasification unit
FLNG – Floating liquefied natural gas
PM – Pierson-Moskowitz
IFFT – Inverse fast Fourier transform
API – American Petroleum Institute
RAO – Response amplitude operator
IEC – International Electro-Technical Commission
HAWT – Horizontal axis wind turbine
VAWT – Vertical axis wind turbine
RNA – Rotor nacelle assembly

Model Exam Paper 3

Maximum marks: 100 **Time: 3 hours**
Answer all questions.
No additional support materials are permitted.

PART A (20 × 1 = 20)

1. _____ are insensitive under lateral loads arising from wind, waves, and currents.
2. A _____ design approach is popularly used to design fixed platforms so they exhibit very low displacements under lateral loads.
3. _____ help restore dynamic equilibrium in the system under various environmental loads.
4. In a spar platform, a _____ generates high buoyancy for the structure.
5. _____, _____, and _____ are passive control devices.
6. Buoyant legs are an alternative structural form of _____ platforms.
7. _____ isolate the deck and buoyant legs in a BLSRP and provide operational comfort.
8. The stress range and the number of cycles are estimated using _____.
9. Random waves are usually described using statistical parameters such as _____ and _____.
10. The average wind velocity occurring over one hour is taken as the _____.
11. A triceratops is _____ in the translational DOF.
12. The peaks in the response spectrum usually occur at _____.
13. Ice floes freezing together results in the formation of _____ covering more than 10 km.
14. _____ controls an ice floe's impact on an offshore structure.
15. Ice failure occurs due to _____ and _____ modes at lower and higher strain rates, respectively.
16. An increase in ice velocity _____ the average crushing force.
17. The principal mechanism that causes a reduction in the strength and stability of a structure during a fire is _____.
18. Young's modulus, stiffness, and the yield strength of structural steel _____ with an increase in temperature.

19. Ring stiffeners prevent damage from spreading to the adjacent bay and act as an obstruction to _____.
20. The suitability of pontoon wind turbines is limited to _____.

PART B (10 × 3 = 30)

1. Differentiate elasticity and recentering.
2. What is a shallow TLD?
3. List the major components in the deck of a BLSRP.
4. Why are rotational responses observed in the deck of a BLSRP, despite the presence of hinged joints?
5. Explain the PM spectrum and its applicability.
6. What is the apparent wave period?
7. What are the special loads that act on offshore structures?
8. Explain the ice–structure interaction phenomenon.
9. List the ice failure modes.
10. What are the advantages of offshore wind turbines?

PART C (5 × 10 = 50)

1. Describe active control systems with a neat sketch.
2. Write briefly about a TLCD.
3. What are the advantages of a TLP with a TMD?
4. Describe the response behavior of a BLSRP under wave loads.
5. Explain the continuous ice crushing phenomenon.

Model Exam Paper 3: KEY

Maximum marks: 100   **Time: 3 hours**
Answer all questions.
No additional support materials are permitted.

PART A (20 × 1 = 20)

1. *Fixed platforms* are insensitive under lateral loads arising from wind, waves, and currents.
2. A *strength-based* design approach is popularly used to design fixed platforms so they exhibit very low displacements under lateral loads.
3. *Large displacements* help restore dynamic equilibrium in the system under various environmental loads.

4. In a spar platform, a *hard tank* generates high buoyancy for the structure.
5. *Tuned mass dampers, tuned liquid dampers,* and *tuned liquid column dampers* are passive control devices.
6. Buoyant legs are an alternative structural form of *spar* platforms.
7. *Hinged joints* isolate the deck and buoyant legs in a BLSRP and provide operational comfort.
8. The stress range and the number of cycles are estimated using *the rainflow-counting method.*
9. Random waves are usually described using statistical parameters such as *significant wave height* and *zero-crossing periods.*
10. The average wind velocity occurring over one hour is taken as the *steady wind velocity*.
11. A triceratops is *monolithic* in the translational DOF.
12. The peaks in the response spectrum usually occur at *multiples or fractions of the natural frequency of the structure or the dominant wave frequency.*
13. Ice floes freezing together results in the formation of *ice fields* covering more than 10 km.
14. *Drift velocity* controls an ice floe's impact on an offshore structure.
15. Ice failure occurs due to *creep* and *crushing* modes at lower and higher strain rates, respectively.
16. An increase in ice velocity *decreases* the average crushing force.
17. The principal mechanism that causes a reduction in the strength and stability of a structure during a fire is *the release of potential energy*.
18. Young's modulus, stiffness, and the yield strength of structural steel *decrease* with an increase in temperature.
19. Ring stiffeners prevent damage from spreading to the adjacent bay and act as an obstruction to *circumferential bending*.
20. The suitability of pontoon wind turbines is limited to *calm seas*.

PART B (10 × 3 = 30)

1. Differentiate elasticity and recentering.
   Elasticity refers to material characteristics and ensures that a member regains its form, shape, and size upon the removal of loads, when the applied load is within the elastic limit. Recentering is an extension of this property.

   Recentering refers to the capability of the structural form (not a material characteristic) to regain its initial position (which may not be an equilibrium position) in the presence of external forces (not upon their removal, unlike in elasticity).

2. What is a shallow TLD?
   If the ratio of the height of the liquid column in a damper to the length of the tank (in the case of a rectangular tank) or the diameter of the circular tank is less than 0.15, then it is classified as a shallow water tuned liquid damper.

3. List the major components in the deck of a BLSRP.
   The deck has utilities including a regasification unit, a gas turbine with a generator, air compressors, fuel pumps, a fire water and foam system, a fresh water system, cranes, a lubrication oil system, lifeboats, a helipad, and a LNG tank.

4. Why are rotational responses observed in the deck of a BLSRP, despite the presence of hinged joints?
   The presence of rotational responses in the deck, despite the presence of hinged joints, is due to the differential heave response that occurs due to dynamic tether tension variations.

5. Explain the PM spectrum and its applicability.
   The most commonly used wave spectrum in offshore design is the PM spectrum, which is applied in different regions such as the Gulf of Mexico, offshore Brazil, Western Australia, offshore Newfoundland, and Western Africa in both operational and survival conditions. This spectrum is suitable for representing open sea conditions that are neither fetch limited nor duration limited.

6. What is the apparent wave period?
   Current in the same direction as waves increases the wavelength and the wave period. The increased wave period (10%) due to the current action is called the apparent wave period.

7. What are the special loads that act on offshore structures?
   - *Environmental loads due to ice, earthquakes, tides, and marine growth*
   - *Loads due to temperature variations and seafloor movement*
   - *Accidental loads due to ship–platform collisions, dropped objects, fires, explosions, changes of intended pressure differences during drilling, and failure of mooring lines in the case of compliant structures*

8. Explain the ice–structure interaction phenomenon.
   When an ice sheet hits a vertical structure under the action of wind, waves, and currents, continuous failure of the ice occurs, which results in a horizontal force on the structure. Under certain conditions, the ice–structure interaction may also result in transient vibrations due to pressure gradients developed from the continuous failure of the ice.

9. List the ice failure modes.
   - *Crushing*
   - *Buckling*
   - *Shear*

- *Radial and circumferential cracking*
- *Creep*
- *Spalling*

10. What are the advantages of offshore wind turbines?
- *Less intense sea turbulence*
- *Fewer constraints on the size of the wind turbines*
- *Avoidance of noise and visual disturbances due to the distance from shore*

PART C (5 × 10 = 50)

1. Describe active control systems with a neat sketch.

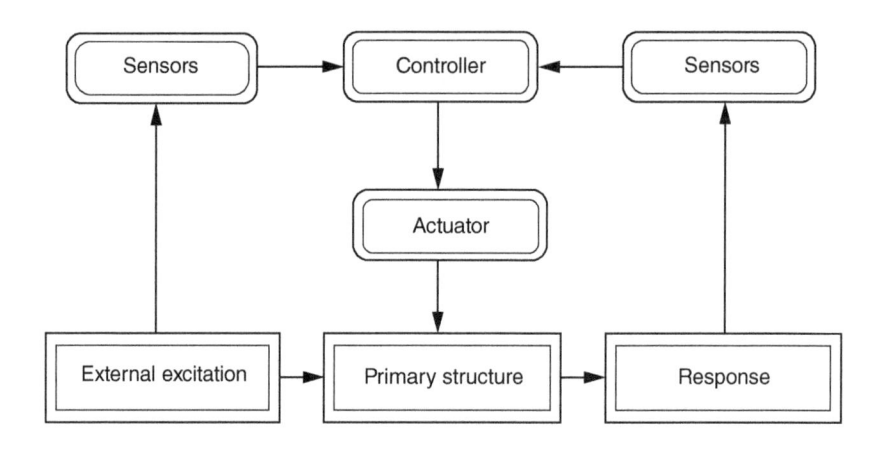

```
┌─────────┐      ┌───────────┐      ┌─────────┐
│ Sensors │─────▶│ Controller│◀─────│ Sensors │
└─────────┘      └───────────┘      └─────────┘
     ▲                 │                  ▲
     │                 ▼                  │
     │           ┌──────────┐             │
     │           │ Actuator │             │
     │           └──────────┘             │
     │                 │                  │
     │                 ▼                  │
┌──────────────┐ ┌──────────────┐  ┌──────────┐
│   External   │▶│   Primary    │─▶│ Response │
│  excitation  │ │  structure   │  │          │
└──────────────┘ └──────────────┘  └──────────┘
```

An active control system consists of three major components:

- *The monitoring system can perceive the present condition of the structure and subsequently record the data using an electronic data acquisition system.*
- *The control system decides what reaction forces to apply to the structure based on communications received from the monitoring system.*
- *The actuating system applies physical forces to the structure as directed by the control system.*

To accomplish all these things, an active control system needs a continuous external power source. The loss of power that might be experienced during a catastrophic event may render these systems ineffective.

Common examples of active control systems are active mass dampers and active liquid dampers.

2. Write briefly about a TLCD.

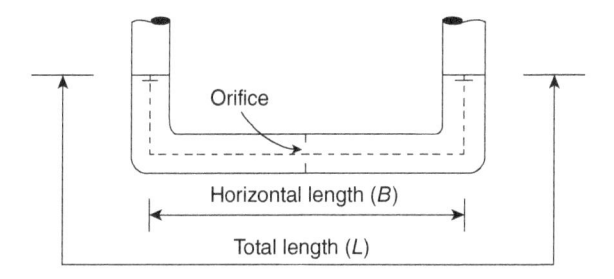

A tuned liquid column damper (TLCD) is a U-shaped tube half-filled with liquid. Unlike a TLD, which depends on liquid sloshing dampening structural vibrations, a TLCD controls structural motion by a combined action of the movement of liquid in the tube and the loss of pressure due to the orifice inside the tube.

A nozzle is placed at the horizontal part of the tube. The extent of response control achieved by a TLCD depends on the frequency of the exciting force acting on the structure.

While the restoring force is developed by the gravitational force acting on the liquid, the orifice is the controlling element for the dynamics of the liquid sloshing inside the tube. Damping depends on the opening and the type of orifice used.

3. What are the advantages of a TLP with a TMD?
   - *A spring-mass system with a higher mass ratio is effective for response reduction with a wide range of time periods.*
   - *A TMD shows better control for larger wave heights.*
   - *An increase in wave elevation increases the surge response at higher periods.*
   - *Adding a TMD to the structure shifts the surge, heave, and pitch natural periods and increases the structure's damping ratio.*
   - *The response reduction is found to be high for the higher mass ratios.*
   - *Greater heave response reduction is observed due to the reduction in the surge response and the tether tension variation.*
   - *By controlling the surge response, indirect control in the heave and pitch DOF is achieved.*

4. Describe the response behavior of a BLSRP under wave loads.
   The deck response is significantly less than the maximum response in all active DOF. It can also be observed that the responses of the deck and buoyant legs are symmetric about the abscissa with less residue indicating high recentering capabilities. This behavior is attributed to the restraint offered by the hinged joints in both the translational and rotational DOF.

Differences in the responses of the buoyant legs are due to the variable submergence effect, which is one of the primary sources of nonlinearity in the excitation force. The presence of hinged joints at each BLS unit isolates the deck from the legs and thus improves the operational comfort and safety of the platform. The presence of rotational responses in the deck, despite the presence of hinged joints, is due to the differential heave response that occurs due to dynamic tether tension variations.

Because the buoyant legs are symmetrically spread with respect to the vertical axis of the platform, it is imperative to envisage a non-uniform phase lag in the recentering process; this causes the yaw motion of the deck. Greater stiffness in the yaw motion is due to the symmetric layout of the buoyant legs, which are spread at the bottom.

A deck response that is significantly less than that of the buoyant legs validates the use of the hinged joints; they do not transfer rotations from the legs to the deck. A lower heave response for the deck, in comparison to that of the BLS units, ensures comfortable and safe operability. The yaw response of the deck is attributed to the time delay in the recentering capability of the buoyant legs under directional wave loads.

5. Explain the continuous ice crushing phenomenon.

   The major factor that limits the maximum ice force acting on any structure is the ice failure mechanism. The ice failure mechanism, in turn, depends upon ice parameters such as the ice thickness, ice velocity, width of the ice plate, and shape of the structure.

   When an ice sheet interacts with a compliant structure, ice failure occurs due to ductile and brittle modes given low and high velocities, respectively. As a result, the continuous ice crushing phenomenon occurs given high ice velocity.

   Ice crushing is a common ice failure mechanism of ice sheets, which results in maximum ice force on structures. It occurs when a sheet of ice hits a vertical-sided structure with moderate to high ice velocity.

   During this process, horizontal cracks form on the ice sheet at the contact zone, leading to pulverization of the ice sheet. The crushed ice particles in the vicinity of the structure pile up and slide around the structure, causing the structure to vibrate.

   The ice forces acting on a structure under crushing ice failure are a function of the ice strength, which depends upon the ice thickness and formation.

   Continuous ice crushing during ice–structure interaction results in non-uniform, partial contact, and non-simultaneous pressure on the contact area.

   The ice force–time history will have waveforms with randomly distributed wave amplitudes and periods. Thus, the ice force can be designated as a stochastic process and described using a frequency spectrum.

   The uncoupled time-dependent load can be used in the dynamic analysis of structures because the transition between the different modes of failure is not completely established.

References

Adrezin, R., Bar-Avi, P., and Benaroya, H. (1996). Dynamic response of compliant offshore structures-review. *Journal of Aerospace Engineering* 9 (4): 114–131.

Agarwal, A.K. and Jain, A.K. (2002). Dynamic behavior of offshore spar platforms under regular sea waves. *Ocean Engineering* 30: 487–516.

Aggarwal, N., Manikandan, R., and Saha, N. (2015). Predicting short term extreme response of spar offshore floating wind turbine. 8th International Conference on Asian and Pacific coast (APAC 2015). *Procedia Engineering* 116: 47–55.

Amdahl, J. and Eberg, E. (1993). Ship collision with offshore structures. *Proceedings of the 2nd European Conference on Structural Dynamics (EURODYN'93)*, Trondheim, Norway (June). EASD. ISBN: 9054103361.

American Bureau of Shipping (2014). LNG bunkering: Technical and operational advisory. Houston: American Bureau of Shipping.

API RP WSD (2005). Recommended practice for planning, designing and constructing fixed offshore platforms-working stress design. Washington, D.C.: American Petroleum Institute.

Balendra, T., Wang, C.M., and Cheong, H.F. (1995). Effectiveness of tuned liquid column dampers for the vibration control of towers. *Engineering Structures* 17 (9): 668–675. https://doi.org/10.1016/0141-0296(95)00036-7.

Bar-Avi, P. (1999). Nonlinear dynamic response of a tension leg platform. *Journal of Offshore Mechanics and Arctic Engineering* 121: 219–226.

Bar-Avi, P. and Benaroya, H. (1996). Non-linear dynamics of an articulated tower in the ocean. *Journal of Sound and Vibration* 190 (1): 77–103.

Bhattacharyya, S.K., Sreekumar, S., and Idichandy, V.G. (2003). Coupled dynamics of sea star mini tension leg platform. *Ocean Engineering* 30: 709–737.

Booton, M., Caswell, R.D., and Tennyson, R.C. (1971). Buckling of imperfect elliptical cylindrical shells under axial compression. *AIAA Journal* 9 (2): 250–255.

References

Booton, M., Joglekar, N., and Deb, M. (1987). The effect of tether damage on tension leg platform dynamics. *Journal of Offshore Mechanics and Arctic Engineering* 109: 186–192.

Brown, D.T. and Mavrakos, S. (1999). Comparative study on mooring line dynamic loading. *Marine Structures* 12 (1999): 131–151.

Buchner, B. and Bunnik, T. (2007). Extreme wave effects on deep water floating structures. Offshore Technology Conference, Houston, Texas (30 April–3 May).

Buchner, B., Wichers, J.E.W., and de Wilde, J.J. (1999). Features of the state-of-the-art deep water offshore basin. *Proceedings of the Offshore Technology Conference*, Houston, Texas (3–6 May).

Butterfield, S., Musial, W., Jonkman, J. et al. (2005). Engineering challenges for floating offshore wind turbines. *Proceedings of the Copenhagen Offshore Wind 2005 Conference and Expedition*, Copenhagen, Denmark (26–28 October). Golden, CO: National Renewable Energy Laboratory.

Capanoglu, C.C., Shaver, C.B., Hirayama, H., and Sao, K. (2002). Comparison of model test results and analytical motion analyses for a buoyant leg structure. *Proceedings of the International Offshore and Polar Engineering Conference*, Kitakyushu, Japan (26–31 May). International Society of Offshore and Polar Engineers.

CEN-EN (2019). Hot finished structural hollow sections. Part 2: Tolerances, dimensions and sectional properties. Brussels, Belgium: European Committee for Standardization (CEN).

Chakrabarti, S.K. (1998). Physical model testing of floating offshore structures. Dynamic Positioning Conference, Houston, Texas (13–14 October).

Chandrasekaran, S. (2014). *Advanced Theory on Offshore Plant FEED Engineering*. Republic of South Korea: Changwon National University Press. ISBN: 978-89-969792-8-9.

Chandrasekaran, S. (2015a). *Dynamic Analysis and Design of Offshore Structures*. Springer, ISBN: 978-81-322-2276-7.

Chandrasekaran, S. (2015b). *Advanced Marine Structures*. Florida: CRC Press. ISBN: 9781498739689.

Chandrasekaran, S. (2016a). *Offshore Structural Engineering: Reliability and Risk Assessment*. Florida: CRC Press. ISBN: 978-149-87-6519-0.

Chandrasekaran, S. (2016b). *Health, Safety and Environmental Management in Offshore and Petroleum Engineering*. Wiley. ISBN: 978-111-92-2184-5.

Chandrasekaran, S. (2017). *Dynamic Analysis and Design of Ocean Structures*, 2e. Singapore: Springer. ISBN:978-981-10-6088-5.

Chandrasekaran, S., Chandak, N.R., and Gupta, A. (2006b). Stability analysis of TLP tethers. *Ocean Engineering* 33: 471–482. https://doi.org/10.1016/j.oceaneng.2005.04.015.

Chandrasekaran, S. and Gaurav, G. (2008). Offshore triangular tension leg platform earthquake motion analysis under distinctly high sea waves. *Ships and Offshore Structures* 3 (3): 173–184. https://doi.org/10.1080/17445300802051681.

Chandrasekaran, S., Gaurav, G., Serino, G., and Miranda, S. (2011). Ringing and springing response of triangular TLPs. *International Shipbuilding Progress* 58: 141–163.

Chandrasekaran, S., Gaurav, S., and Jain, A.K. (2010). Ringing response of offshore compliant structures. *International Journal of Ocean and Climate Systems* 1 (3 & 4): 133–143.

Chandrasekaran, S. and Jain, A.K. (2002a). Dynamic behavior of square and triangular offshore tension leg platforms under regular wave loads. *Ocean Engineering* 29 (3): 279–313. https://doi.org/10.1016/S0029-8018(00)00076-7.

Chandrasekaran, S. and Jain, A.K. (2002b). Triangular configuration tension leg platform behaviour under random sea wave loads. *Ocean Engineering* 29: 1895–1928.

Chandrasekaran, S. and Jain, A.K. (2004). Aerodynamic behavior of offshore triangular tension leg platforms. *Proceedings of the ISOPE*, Toulon, France (23–28 May). International Society of Offshore and Polar Engineers.

Chandrasekaran, S. and Jain, A.K. (2016). *Ocean Structures: Construction, Materials and Operations*. Florida: CRC Press, ISBN: 978-149-87-9742-9.

Chandrasekaran, S., Jain, A.K., and Chandak, N.R. (2004). Influence of hydrodynamic coefficients in the response behavior of triangular TLPs in regular waves. *Ocean Engineering* 31: 2319–2342.

Chandrasekaran, S., Jain, A.K., and Chandak, N.R. (2006a). Seismic analysis of offshore triangular tension leg platforms. *International Journal of Structural Stability and Dynamics* 6 (1): 97–120.

Chandrasekaran, S., Jain, A.K., Gupta, A., and Srivastava, A. (2007b). Response behavior of triangular tension leg platforms under impact loading. *Ocean Engineering* 34: 45–53.

Chandrasekaran, S., Jain, A.K., Gupta, A., and Srivastava, A. (2007b). Response behavior of triangular tension leg platforms under impact loading. *Ocean Engineering* 34: 45–53.

Chandrasekaran, S., Jain, A.K., and Gupta, A. (2007a). Influence of wave approach angle on TLP's response. *Ocean Engineering* 34: 1322–1327. https://doi.org/10.1016/j.oceaneng.2006.08.007.

Chandrasekaran, S. and Kiran, P.A. (2018). Mathieu stability of offshore triceratops under postulated failure. *Ships and Offshore structures* 13 (2): 143–148. https://doi.org/10.1080/17445302.2017.133578.

Chandrasekaran, S. and Lognath, R.S. (2015). Dynamic analyses of buoyant leg storage regasification platform (BLSRP) under regular waves: experimental investigations. *Ships and Offshore Structures* 12 (2): 227–232.

Chandrasekaran, S. and Madhuri, S. (2015). Dynamic response of offshore triceratops: numerical and experimental investigations. *Ocean Engineering* 109: 401–409.

Chandrasekaran, S., Madhuri, S., and Jain, A.K. (2013). Aerodynamic response of offshore triceratops. *Ships and Offshore Structures* 8 (2): 123–140.

Chandrasekaran, S. and Mayanak, S. (2017). Dynamic analyses of stiffened triceratops under regular waves: experimental investigations. *Ships and Offshore Structures* 12 (5): 697–705. https://doi.org/10.1080/17445302.2016.1200957.

Chandrasekaran, S., Mayank, S., and Jain, A. (2015). Dynamic response behavior of stiffened triceratops under regular waves: Experimental investigations. *Proceedings of the 34th International Conference on Ocean, Offshore and Arctic Engineering (OMAE 2015)*, St. John's, NL, Canada (31 May–5 June). ASME.

Chandrasekaran, S. and Nagavinothini, R. (2017). Analysis and design of offshore triceratops under ultra-deep waters. *International Journal of Civil, Environmental, Structural, Construction and Architectural Engineering* 11 (11): 1520–1528.

Chandrasekaran, S. and Nagavinothini, R. (2018a). Dynamic analyses and preliminary design of offshore triceratops in ultra-deep waters. *International Journal of Innovative Infrastructure Solutions* 3 (1): 16. https://doi.org/10.1007/s41062-017-0124-1.

Chandrasekaran, S. and Nagavinothini, R. (2018b). Tether analyses of offshore triceratops under wind, wave and current. *Marine Systems & Ocean Technology* 13: 34–42. https://doi.org/10.1007/s40868-018-0043-9.

Chandrasekaran, S. and Nagavinothini, R. (2019a). Tether analyses of offshore triceratops under ice loads due to continuous crushing. *International Journal of Innovative Infrastructure Solutions* 4: 25. https://doi.org/10.1007/s41062-019-0212-5.

Chandrasekaran, S. and Nagavinothini, R. (2019b). Ice-induced response of offshore triceratops. *Ocean Engineering* 180: 71–96. https://doi.org/10.1016/j.oceaneng.2019.03.063.

Chandrasekaran, S. and Seeram, M. (2012). Stability studies on offshore triceratops. *Intl J of Research & Development* 1 (10): 398–404.

Chandrasekaran, S., Sharma, A., and Srivastava, S. (2007c). Offshore triangular TLP behavior using dynamic Morison equation. *Journal of Structural Engineering* 34 (4): 291–296.

Chandrasekaran, S., Sundaravadivelu, R., Pannerselvam, R., and Madhuri, S. (2011). Experimental investigations of offshore triceratops under regular waves. *Proceedings of the 30th International Conference on Ocean, Offshore and Arctic Engineering (OMAE 2015)*, Rotterdam, The Netherlands (19–24 June). ASME.

Chandrasekaran, S., Thaillammai, C., and Khader, S.A. (2016). Structural health monitoring of offshore structures using wireless sensor networking under operational and environmental variability. *International Journal of Environmental, Chemical and Ecological Engineering* 10 (1): 33–39.

Chatjigeorgiou, I.K. and Mavrakos, S.A. (2010). An analytical approach for the solution of the hydrodynamic diffraction by arrays of elliptical cylinders. *Applied Ocean Research* 32 (2): 242–251.

Cho, S.R., Choi, S.I., and Son, S.K. (2015). Dynamic material properties of marine steels under impact loadings. *Proceedings of the 2015 World Congress on Advances in Structural Engineering and Mechanics*, Incheon, Korea. IASEM.

Chou, F.S.F. (1980). Analytical approach to the design of a tension leg platform. *Proceedings of the Offshore Technology Conference*, Houston, Texas (5–8 May).

ClassNK (2015). Guidelines for floating offshore facilities for LNG production, storage, offloading and regasification. Tokyo, Japan.

Colwell, S. and Basu, B. (2009). Tuned liquid column dampers in offshore wind turbines for structural control. *Engineering Structures* 31 (2): 358–368. https://doi.org/10.1016/j.engstruct.2008.09.00.

Copple, R.W. and Capanoglu, C.C. (1995). A buoyant leg structure for the development of marginal fields in deep water. *Proceedings of the 5th International Offshore and Polar Engineering Conference*, The Hague, The Netherlands (11–16 June). International Society of Offshore and Polar Engineers.

Davenport, A.G. (1961). The application of statistical concepts to the wind loading of structures. *Proceedings Institution of Civil Engineers* 19: 449–471.

de Boom, W.C., Pinkster, J.A., and Tan, P.S.G. (1984). Motion and tether force prediction of a TLP. *Journal of Waterway, Port, Coastal and Ocean Engineering* 110 (4): 472–486.

Den Hartog, J.P. (1985). *Mechanical Vibrations*. NY: Dover publications Inc.

Det Norske Veritas (2010a). Fatigue design of offshore steel structures. Recommended Practice DNV-RP-C203.

Det Norske Veritas (2010b). Design against accidental loads. Recommended Practice DNV-RP-C204.

Det Norske Veritas (2011). Floating liquefied gas terminal offshore technical guidance 2. Oslo, Norway.

Do, Q.T., Muttaqie, T., Shin, H.K., and Cho, S.R. (2018). Dynamic lateral mass impact on steel stringer-stiffened cylinders. *International Journal of Impact Engineering* 116: 105–126. https://doi.org/10.1016/j.ijimpeng.2018.02.007.

Donely, M.G. and Spanos, P.D. (1991). Stochastic response of a tension leg platform to viscous drift forces. *Journal of offshore Mechanics and Arctic Engineering* 113: 148–155.

El-gamal, A.R., Essa, A., and Ismail, A. (2013). Effect of tethers tension force in the behavior of a tension leg platform subjected to hydrodynamic force. *International Journal of Civil, Structural, Construction and Architectural Engineering* 7 (12): 645–652.

Ertas, A. and Ekwaro-Osire, S. (1991). Effect of damping and wave parameters on offshore structure under random excitation. *Nonlinear Dynamics* 2 (2): 119–136. https://doi.org/10.1007/BF00053832.

Ertas, A. and Lee, J.-H. (1989). Stochastic response of tension leg platform to wave and current forces. *Journal of Energy Resources Technology* 111: 221–230.

Farshidianfar, A., Oliazadeh, P., and Farivar, H.R. (2009). Optimal parameter's design in tuned liquid column damper. 17th Annual International Conference on Mechanical Engineering, University of Tehran, Iran.

Finn, L.D., Maher, J.V., and Gupta, H. (2003). The cell spar and vortex induced vibrations. *Proceedings of the Offshore Technology Conference*, Houston, Texas (5–8 May).

Fujino, Y. and Abe, M. (1993). Design formulas for tuned mass dampers based on a perturbation technique. *Earthquake Engineering and Structural Dynamics* 22 (10): 833–854. https://doi.org/10.1002/eqe.4290221002.

Gao, H., Kwok, K.C.S., and Samali, B. (1997). Optimization of tuned liquid column damper. *Engineering Structures* 19 (6): 476–486. https://doi.org/10.1016/S0141-0296(96)00099-5.

Gasim, M.A., Kurian, V.J., Narayanan, S.P., and Kalaikumar, V. (2008). Responses of square and triangular TLPs subjected to random waves. International conference on construction and building technology, Universiti Teknologi Petronas, Malaysia (16–20 June).

Glanville, R.S., Paulling, J.R., Halkyard, J.E., and Lehtinen, T.J. (1991). Analysis of the spar floating, drilling, production and storage structure. *Proceedings of the Offshore Technology Conference*, Houston, Texas (6–9 May).

Graham, R.P. and Webb, R.M. (1980). Tethered buoyant platform production system. *Proceedings of the Offshore Technology Conference*, Houston, Texas (5–8 May).

Gruben, G., Langseth, M., Fagerholt, E., and Hopperstad, O.S. (2016). Low-velocity impact on high-strength steel sheets: an experimental and numerical study. *International Journal of Impact Engineering* 88: 153–171. https://doi.org/10.1016/j.ijimpeng.2015.10.001.

Halkyard J.E., Davies, R.L., and Glanville, R.S. (1991). The tension buoyant tower: A design for deep water, *Proceedings of the 3rd Annual Offshore Technology Conference*, Houston, Texas (6–9 May).

Harding, J.E., Onoufriou, A., and Tsang, S.K. (1983). Collisions – what is the danger to offshore rigs. *Journal of Constructional Steel Research* 3 (2): 31–38. https://doi.org/10.1016/0143-974X(83)90020-2.

Heinonen, J. and Rissanen, S. (2017). Coupled-crushing analysis of a sea ice-wind turbine interaction – feasibility study of FAST simulation software. *Ships and Offshore Structures* 12 (8): 1056–1063. https://doi.org/10.1080/17445302.2017.1308782.

Hwang, J.-K., Roh, M.-I., and Lee, K.-Y. (2010). Detailed design and construction of the hull of a floating, production, storage and off-loading (FPSO) unit. *Ships and Offshore Structures* 5 (2): 93–104. https://doi.org/10.1080/17445300903169168.

IEC (2005). Wind turbines – Part 1: Design requirements. 61400–1 Ed. 3.

Infanti, S., Robinson, J., and Smith, R. (2008). Viscous dampers for high-rise buildings. 14th World Conference on Earthquake Engineering, Beijing, China (12–17 October).

Islam, N. and Ahmad, S. (2003). Nonlinear seismic response of articulated offshore tower. *Defence Science Journal* 53 (1): 105–113.

Jain, A.K. (1997). Nonlinear coupled response of offshore TLP to regular waves. *Ocean Engineering* 24 (7): 577–592.

Jain A.K. and Chandrasekaran, S. (2004). Aerodynamic behavior of offshore triangular tension leg platforms. *Proceedings of the 14th International Offshore and Polar Engineering Conference*, Toulon, France (23–28 May). International Society of Offshore and Polar Engineers.

Jayalekshmi, R., Sundaravadivelu, R., and Idichandy, V.G. (2010). Dynamic analysis of deep water tension leg platforms under random waves. *Journal of Offshore Mechanics and Arctic Engineering* 132: 041605-1–041605-4.

Jefferys, E.R. and Patel, M.H. (1982). Dynamic analysis models of tension leg platforms. *Journal of Energy Resources Technology* 104: 217–223.

Jin, Q., Li, X., Sun, N. et al. (2007). Experimental and numerical study on tuned liquid dampers for controlling earthquake response of jacket offshore platform. *Marine Structures* 20 (4): 238–254. https://doi.org/10.1016/j.marstruc.2007.05.002.

Jumppanen, P. (1984). Structural engineering in arctic regions. IABSE congress report.

Kareem, A. and Sun, W.J. (1987). Stochastic response of structures with fluid-containing appendages. *Journal of Sound and Vibration* 119 (3): 389–408. https://doi.org/10.1016/0022-460X(87)90405-6.

Kareem, A. (1983). Mitigation of wind induced motion of tall buildings. *Journal of Wind Engineering and Industrial Aerodynamics* 11: 273–284. https://doi.org/10.1016/0167-6105(83)90106-X.

Kareem, A. (1990). Reduction of wind induced motion utilizing a tuned sloshing damper. *Journal of Wind Engineering and Industrial Aerodynamics* 36: 725–737. https://doi.org/10.1016/0167-6105(90)90070-S.

Karna, T., Qu, Y., Bi, X. et al. (2007). A spectral model for forces due to ice crushing. *Journal of Offshore Mechanics and Arctic Engineering* 129 (2): 138–145. https://doi.org/10.1115/1.2426997.

Karna, T., Qu, Y., and Yue, Q. (2006a). An equivalent lateral force for continuous crushing. In: *Proceedings of the 25th International Conference on Offshore Mechanics and Arctic Engineering*. ASME https://doi.org/10.1115/OMAE2006-92648.

Karna, T., Qu, Y., and Yue, Q.J. (2006b). Baltic model of global ice forces on vertical structures. *Proceedings of the 18th IAHR International Symposium on Ice*.

Karr, D.G., Troesch, A.W., and Wingate, W.C. (1993). Nonlinear dynamic response of a simple ice-structure interaction model. *Journal of Offshore Mechanics and Arctic Engineering* 115 (4): 246–252. https://doi.org/10.1115/1.2920119.

References

Kim, C.-H., Lee, C.-H.O., and Goo, J.-S. (2007). A dynamic response analysis of tension leg platforms including hydrodynamic interaction in regular waves. *Ocean Engineering* 34: 1680–1689.

Kim, K.J., Lee, J.H., Park, D.K. et al. (2016). An experimental and numerical study on nonlinear impact responses of steel-plated structures in an Arctic environment. *International Journal of Impact Engineering* 93: 99–115. https://doi.org/10.1016/j.ijimpeng.2016.02.013.

Kobayashi, M., Shimada, K., and Fujihira, T. (1987). Study on dynamic responses of a TLP in waves. *Journal of Offshore Mechanics and Arctic Engineering* 109: 61–66.

Koo, B.J., Kim, M.H., and Randall, R.E. (2004). Mathieu instability of a spar platform with mooring and risers. *Ocean Engineering* 31: 2175–2208.

Kurian, V.J., Gasim, M.A., Narayan, S.P., and Kalaikumar, V. (2008). Parametric study of TLPs subjected to random waves. *Proceedings of the International Conference on Construction and Building Technology, Conference C (International conference on Structural Engineering)*, Kuala Lumpur, Malaysia, (16–20 June).

Kurian, V.J., Idichandy, V.G., and Ganapathy, C. (1993). Hydro dynamic response of tension-leg platforms: A model. *Experimental Mechanics*: 212–217.

Lee, C.-L., Chen, Y.-T., Chung, L.-L., and Wang, Y.-P. (2006). Optimal design theories and applications of tuned mass dampers. *Engineering Structures* 28 (1): 43–53. https://doi.org/10.1016/j.engstruct.2005.06.023.

Lee, H.H., Wong, S.H., and Lee, R.S. (2006). Response mitigation on the offshore floating platform system with tuned liquid column damper. *Ocean Engineering* 33: 1118–1142. https://doi.org/10.1016/j.oceaneng.2005.06.008.

Lee, H.H. and Juang, H.H. (2012). Experimental study on the vibration mitigation of offshore tension leg platform system with UWTLCD. *Smart Structures and Systems* 9 (1): 71–104.

Leonard, J.W. and Young, R.A. (1985). Coupled response of compliant offshore platforms. *Engineering Structures* 7: 21–31.

Liagre, P.F. and Niedzwecki, J.M. (2003). Estimating nonlinear coupled frequency-dependent parameters in offshore engineering. *Applied Ocean Research* 25: 1–19.

Lloyd's Register (2005). Rules and regulations for the classification of mobile offshore units. London, U.K.

Logan, B.L., Naylor, S., Munkejord, T., and Nyhgaard, C. (1996). Atlantic alliance: The next generation tension leg platform. *Proceedings of the Offshore Technology Conference*.

Low, Y.M. (2009). Frequency domain analysis of a tension leg platform with statistical linearization of the tendon restoring forces. *Marine Structures* 22: 480–503.

Mahilvahanan, A.C. and Selvam, R.P. (2010). Static and dynamic analysis of semi-submersible type floater for offshore wind turbine. *Proceedings of MARTEC*.

Mahoney, A., Eicken, H., Shapiro, L., and Grenfell, T.C. (2004). Ice motion and driving forces during a spring ice shove on the Alaskan Chukchi coast. *Journal of Glaciology* 50 (169): 195–207. https://doi.org/10.3189/172756504781830141.

Manco, M.R., Vaz, M.A., Cyrino, J.C., and Landesmann, A. (2013). Behavior of stiffened panels exposed to fire. *Proceedings of IV MARSTRUCT*, Espoo, Finland. ISBN 978-1-138-00045-2.

Matha, D. (2009). Model development and loads analysis of an offshore wind turbine on a tension leg platform, with a comparison to other floating turbine concepts. Report NREL/SR-500-45891, National Renewable Energy Laboratory, USA.

McCoy, T.J., Brown, T., and Byrne, A. (2014). Ice load project final technical report (No. DDRP0133). Seattle, WA: DNV GL. https://doi.org/10.2172/1303304.

McGovern, D.J. and Bai, W. (2014). Experimental study on kinematics of sea ice floes in regular waves. *Cold Regions Science and Technology* 103: 15–30. https://doi.org/10.1016/j.coldregions.2014.03.004.

Mekha, B.B., Johnson, C.P., and Roesset, J.M. (1996). Implication of tendon modeling on nonlinear response of TLP. *Journal of Structural Engineering* 122 (2): 142–149.

Mercier R.S., Schott, W.E., Howell, C.T. et al. (1997). Mars tension leg platform - Use of scale model testing in the global design. *Proceedings of the Offshore Technology Conference*, Houston, Texas (5–8 May).

Moharrami, M. and Tootkaboni, M. (2014). Reducing response of offshore platforms to wave loads using hydrodynamic buoyant mass dampers. *Engineering Structures* 81: 162–174. https://doi.org/10.1016/j.engstruct.2014.09.037.

Montasir, O.A. and Kurian, V.J. (2011). Effect of slowly varying drift forces on the motion characteristics of truss spar platforms. *Ocean Engineering* 38: 1417–1429.

Montasir, O.A.A., Kurian, V.J., Narayanan, S.P., and Mubarak, M.A.W. (2008). Dynamic response of spar platforms subjected to waves and current. International Conference on Construction and Building Technology, ICCBT 2008, Kuala Lumpur, Malaysia (16–20 June).

Muren, J., Flugstad, P., Greiner, B. et al. (1996). The 3 column TLP-A cost efficient deep water production and drilling platform. *Proceedings of the Offshore Technology Conference*.

Murray, J.J. and Mercier, R.S. (1996). Model tests on a tension leg platform using truncated tendons. In: *Proceedings of the Workshop on Model Testing of Deep Sea Offshore Structures*. IITC.

Musial, W. and Butterfield, S. (2006). Future for offshore wind energy in the United States. *Proceedings of EnergyOcean*, Palm Beach, Florida (June 2004). Golden, CO: National Renewable Energy Laboratory.

Musial, W., Butterfield, S., and Ram, B. (2006). Energy from offshore wind. Offshore Technology Conference.

Nagamani, K. and Ganapathy, C. (2000). The dynamic response of a three leg articulate tower. *Ocean Engineering* 27: 1455–1471.

Newman, J.N. (1963). The motions of spar buoy in regular waves. Report 1499, David Taylor Model Basin.

Niedzweki, J.M., van de Lindt, J.W., Gage, J.H., and Teigen, P.S. (2000). Design estimates of surface wave interaction with compliant deepwater platforms. *Ocean Engineering* 27: 867–888.

Nielsen, F.G. and Bindingbø, A.U. (2000). Extreme loads in taut mooring lines and mooring line induced damping: an asymptotic approach. *Applied Ocean Research* 22 (2000): 103–118.

Nordgren, R.P. (1987). Analysis of high frequency vibration of tension leg platforms. *Journal of Offshore Mechanics and Arctic Engineering* 109: 119–125.

O'Kane, J.J., Troeschand, A.W., and Thiagaraja, K.P. (2002). Hull component interaction and scaling for TLP hydrodynamic coefficients. *Ocean Engineering* 29: 513–532.

Paik, J.K., Sohn, J.M., Shin, Y.S., and Suh, Y.S. (2011). Nonlinear structural analysis of membrane-type LNG carrier cargo containment system under cargo static pressure loads at the cryogenic condition with a temperature of −163°C. *Ships and Offshore Structures* 6 (4): 311–322. https://doi.org/10.1080/17445302.2010.530428.

Pall, A., Vezina, S., Proulx, P., and Pall, R. (1993). Friction-dampers for seismic control of Canadian space agency headquarters. *Earthquake Spectra* 9 (3): 547–557. https://doi.org/10.1193/1.1585729.

Patel, M.H. and Lynch, E.J. (1983). Coupled dynamics of tensioned buoyant platforms and mooring tethers. *Engineering Structures* 5 (4): 2099–2308. https://doi.org/10.1016/0141-0296(83)90009-3.

Patel, M.H. and Park, H.I. (1991). Dynamics of tension leg platform tethers at Low tension. Part I – Mathieu stability at large parameters. *Marine Structures* 4 (3): 257–273.

Perryman, S.R., Horton, E.E., and Halkyard, J.E. (1995). Tension buoyant tower for small fields in deep waters. *Proceedings of the Offshore Technology Conference*, Houston, Texas (1–4 May).

Ramachandran, G.K.V., Bredmose, H., Sørensen, J.N., and Jensen, J.J. (2014). Fully coupled three- dimensional dynamic response of a tension-leg platform floating wind turbine in waves and wind. *Journal of Offshore Mechanics and Arctic Engineering* 136: 020901–020901.

Ran, Z., Kim, M.H., Niedzwecki, J.M., and Johnson, R.P. (1996). Response of a spar platform in random waves and currents (experiments vs. theory). *International Journal of Offshore Polar Engineering* 6 (1).

Rana, R. and Soong, T.T. (1998). Parametric study and simplified design of tuned mass dampers. *Engineering Structures* 20 (3): 193–204. https://doi.org/10.1016/S0141-0296(97)00078-3.

Ranjani, R. (2015). Response control of tension leg platform using tuned mass damper. PhD thesis. IIT Madras, India.

Reddy, D.V. and Swamidas, A.S.J. (2016). *Essentials of Offshore Structures: Framed and Gravity Platforms*. CRC press ISBN: 9781482220186.

Rho, J.B., Choi, H.S., Lee, W.C. et al. (2002). Heave and pitch motion of a spar platform with damping plate. *Proceedings of the 12th International Offshore and Polar Engineering Conference*, Kitakyshu. International Society of Offshore and Polar Engineers.

Rho, J.B., Choi, H.S., Lee, W.C. et al. (2003). An experimental study for mooring effects on the stability of spar platform. *Proceedings of the 13th International Offshore and Polar Engineering Conference*, Honolulu, Hawaii. International Society of Offshore and Polar Engineers.

Rivera, M.R.M., Vaz, M.A., Cyrino, J.C.R., and Landesmann, A. (2014). Analysis of oil tanker deck under hydrocarbon fire. *International Journal of Modeling and Simulation for the Petroleum Industry* 8 (2): 17–24.

Roitman, N., Andrade, R.F.M., and Batista, R.C. (1992). Dynamic response analysis of small scale model tension leg platform. *Marine Structures* 5: 491–513.

Sadek, F., Mohraz, B., and Lew, H.S. (1998). Single and multiple tuned liquid column dampers for seismic applications. *Earthquake Engineering and Structural Dynamics* 27: 439–463.

Schwartz, M.L. (2005). *Encyclopaedia of Coastal Science*. Netherlands: Springer.

Sellers, L.L. and Niedzwecki, J.M. (1992). Response characteristics of multi-articulated offshore towers. *Ocean Engineering* 19 (1): 1–20.

Shaver, C.B., Capanoglu, C.C., and Serrahn, C.S. (2001). Buoyant leg structures: Preliminary design, constructed cost and model test results. *Proceedings of the Eleventh International Offshore and Polar Engineering Conference*, Stavanger, Norway (17–22 June). International Society of Offshore and Polar Engineers.

Sheng, D., Huajun, L., Ming, L., and Takayama, T. (2002). Experimental study on the effectiveness of TLDs under wave loading. *Journal of Ocean University of Qingdao* 1 (1): 80–86. https://doi.org/10.1007/s11802-002-0036-2.

Shih, L.Y. (1991). Analysis of ice-induced vibrations on a flexible structure. *Applied Mathematical Modelling* 15 (11–12): 632–638. https://doi.org/10.1016/S0307-904X(09)81009-3.

Silvestre, N. (2008). Buckling behaviour of elliptical cylindrical shells and tubes under compression. *International Journal of Solids and Structures* 45 (16): 4427–4447.

Simiu, E. and Leigh, S.D. (1984). Turbulent wind and tension leg platform surge. *Journal of Structural Engineering* 110 (4): 785–802.

Simos, A.N. and Pesce, C.P. (1997). Mathieu stability in the dynamics of TLP tether considering variable tension along the length. *Transactions on Built Environment* 29: 175–186.

Sodhi, D.S. and Haehnel, R.B. (2003). Crushing ice forces on structures. *Journal of Cold Regions Engineering* 17 (4): 153–170. https://doi.org/10.1061/(ASCE)0887-381X(2003)17:4(153).

Sohn, Y., Kim, S., and Yoon, I. (2012). Conceptual design of LNG FSRU topside regasification plant. *Proceedings of the 22nd International Offshore and Polar Engineering Conference*, Rhodes, Greece (17–22 June). International Society of Offshore and Polar Engineers. ISBN 978-1-880653-94-4 (set).

Spanos, P.D. and Agarwal, V.K. (1984). Response of a simple tension leg platform model to wave forces calculated at displaced position. *Journal of Energy Resources Technology* 106 (4): 437–443.

Stansberg, C.T., Karlsen, S.I., Ward, E.G. et al. (2004). Model testing for ultra-deep waters. *Proceedings of the Offshore Technology Conference*, Houston, Texas.

Stansberg, C.T., Ormberg, H., and Oritsland, O. (2002). Challenges in deep water experiments: hybrid approach. *Journal of Offshore Mechanics and Arctic Engineering* 124: 90–96.

Sun, L.M., Fujino, Y., and Koga, K. (1995). A model of tuned liquid damper for suppressing pitching motions of structures. *Earthquake Engineering and Structural Dynamics* 24 (5): 625–636. https://doi.org/10.1002/eqe.4290240502.

Sun, S. and Shen, H.H. (2012). Simulation of pancake ice load on a circular cylinder in a wave and current field. *Cold Regions Science and Technology* 78: 31–39. https://doi.org/10.1016/j.coldregions.2012.02.003.

Syngellakis, S. and Balaji, R. (1989). Tension leg platform response to impact forces. *Marine Structures* 2 (2): 151–171. https://doi.org/10.1016/0951-8339(89)90010-5.

Tabeshpour, M.R., Golafshani, A.A., and Seif, M.S. (2006). Comprehensive study on the results of tension leg platform responses in random sea. *Journal of Zhejiang University Science A* 7 (8): 1305–1317.

Taflanidis, A.A., Scruggs, J.T., and Angelides, D.C. (2008). Robust design optimization of mass dampers for control of tension leg platforms. *Proceedings of the Eighteenth International Offshore and Polar Engineering Conference*, Vancouver, BC, Canada (6–11 July). International Society of Offshore and Polar Engineers.

Taflanidis, A.A., Angelides, D.C., and Scruggs, J.T. (2009). Simulation-based robust design of mass dampers for response mitigation of tension leg platforms. *Engineering Structures* 31 (4): 847–857. https://doi.org/10.1016/j.engstruct.2008.11.014.

Tait, M.J., Isyumov, N., and El Damatty, A. (2008). Performance of tuned liquid dampers. *Journal of Engineering Mechanics* 134 (5): 417–427. https://doi.org/10.1061/(ASCE)0733-9399(2008)134:5(417).

Thiagarajan, K.P. and Troesch, A.W. (1998). Effects of appendages and small currents on the hydro dynamic heave damping of TLP columns. *Journal of Offshore Mechanics and Arctic Engineering* 120: 37–42.

Tigli, O.F. (2012). Optimum vibration absorber (tuned mass damper) design for linear damped systems subjected to random loads. *Journal of Sound and Vibration* 331 (13): 3035–3049. https://doi.org/10.1016/j.jsv.2012.02.017.

Vannucci, P. (1996). Simplified optimal design of tension leg platform, structural. *Optimization* 12: 265–268.

Vickery, P.J. (1990). Wind & wave loads on a tension leg platform: theory and experiment. *Journal of Wind Engineering and Industrial Aerodynamics* 36: 905–914.

Viet, L.D. and Nghi, N.B. (2014). On a nonlinear single-mass two-frequency pendulum tuned mass damper to reduce horizontal vibration. *Engineering Structures* 81: 175–180. https://doi.org/10.1016/j.engstruct.2014.09.03.

Wang, L. and Isberg, J. (2015). Nonlinear passive control of a wave energy converter subject to constraints in irregular waves. *Energies* 8 (7): 6528–6542. https://doi.org/10.3390/en8076528.

Wang, P., Zhao, M., Du, X., and Liu, J. (2019). Analytical solution for the short-crested wave diffraction by an elliptical cylinder. *European Journal of Mechanics-B/Fluids* 74: 399–409.

White, C.N., Copple, R.W., and Capanoglu, C. (2005). Triceratops: An effective platform for developing oil and gas fields in deep and ultradeep water. *Proceedings of the 15th International Offshore and Polar Engineering Conference*. International Society of Offshore and Polar Engineers.

Witz, J.A., Patel, M.H., and Harrison, J.H. (1986). On the hydrodynamics of semisubmersibles with articulated members. *Proceedings of Royal Society of London, Series A, Mathematical and Physical Sciences* 403: 81–109.

Wong, K. (2008). Seismic energy dissipation of inelastic structures with tuned mass dampers. *Journal of Engineering Mechanics* 134 (2): 162–172. https://doi.org/10.1061/(ASCE)0733-9399(2008)134:2(163).

World Meteorological Organization (2014). Sea state code. Geneva, Switzerland.

Wu, H.-l., Chen, X.-j., Huang, Y.-x., and Wang, B. (2014). Influence of the legs underwater on the hydrodynamic response of the multi-leg floating structures. *Ships and Offshore Structures* 9 (6): 578–595.

Wu, J.-C., Shih, M.H., Lin, Y.-Y., and Shen, Y.C. (2005). Design guidelines for tuned liquid column damper for structures responding to wind. *Engineering Structures* 27 (13): 1893–1905. https://doi.org/10.1016/j.engstruct.2005.05.009.

Yan, F.-s., Da-gang, Z., Li-Ping, S., and Yang-shan, D. (2009). Stress verification of a TLP under extreme wave environment. *Journal of Marine Science Applications* 8: 132–136.

Yashima, N. (1976). Experimental and theoretical studies of a tension leg platform in deep water. *Proceedings of the Offshore Technology Conference*, Houston, Texas (3–6 May).

Yoneya, T. and Yoshida, K. (1982). Dynamics of tension leg platforms in waves. *Journal of Energy Resources Technology* 104: 20–28.

Yoshida, K., Ozaki, M., and Oka, N. (1984). Structural response analysis of tension leg platforms. *Journal of Energy Resources* 106: 10–17.

Younis, B.A., Teigen, P., and Przulj, V.P. (2001). Estimating the hydrodynamic forces on a mini TLP with computational fluid dynamics and design-code techniques. *Ocean Engineering* 28: 585–602.

Yue, Q., Bi, X., Zhang, X., and Karna, T. (2002). Dynamic ice forces caused by crushing failure. *Proceedings of the 16th IAHR Symposium on Ice*, Dunedin, New Zealand (December).

Yue, Q., Zhang, X., Bi, X., and Shi, Z. (2001). Measurements and analysis of ice induced steady state vibration. *Proceedings of the International Conference on Port and Ocean Engineering Under Arctic Conditions*.

Zeng, X.-h., Shen, X.-p., and Wu, Y.-x. (2007). Governing equations and numerical solutions of tension leg platform with finite amplitude motion. *Applied Mathematics and Mechanics (English Edition)* 28 (1): 37–49.

Zhang, F., Yang, J., Li, R., and Chen, G. (2007). Numerical investigation on the hydrodynamic performances of a new spar concept. *Science Direct Journal of Hydrodynamics* 19 (4): 473–481.

Zhao, W.H., Yang, J.M., Hu, Z.Q., and Wei, Y.F. (2011). Recent developments on the hydrodynamics of floating liquid natural gas (FLNG). *Ocean Engineering* 38 (2011): 1555–1567.

Zhao, W., Yang, J., and Hu, Z. (2013). Effects of sloshing on the global motion responses of FLNG. *Ships and Offshore Structures* 8 (2): 111–122. https://doi.org/10.1080/17445302.2012.691272.

Ziemer, G. and Evers, K.U. (2016). Model tests with a compliant cylindrical structure to investigate ice-induced vibrations. *Journal of Offshore Mechanics and Arctic Engineering* 138 (4): 041501. https://doi.org/10.1115/1.4033712.

Index

a

accelerometers   65
active control algorithm   25
Airy wave theory   13, 15–17, 23, 97
American Bureau of Shipping (ABS) rules   62
apparent wave period   101
articulated tower (AT)
   BLSRP, 60
   guyed towers   19–21
   offshore compliant platforms   44–48
   response control of   48–53

b

ball joints   18, 21, 139, 183
bergy bits   126
BLSRP *see* buoyant leg storage and regasification platform (BLSRP)
Borgman's method   21
buckling, ice failure modes   125
buoyant leg responses
   characteristics of   154
   deck and   139–140
   with high sea conditions   120
   of wind, waves, and current   120–121

buoyant leg storage and regasification platform (BLSRP)
   articulated towers   60
   buoyant leg structures   62–63
   critical observations   76–85
   experimental investigations   65–72
   experimental setup   64–65
   fatigue analysis of   90–93
   floating production, storage, and offloading   63–64
   numerical studies   72–73
   stability analysis of   85–90
   wave heading angle   69
buoyant leg structure (BLS)   18, 61–63

c

caisson effect   73
capital expenditures (CAPEX)   2
Chakrabarti's stretching formula   23
collision speed   148
collision zone location   152–153
compliance   2
compliant offshore structures   2, 11, 12, 19, 36–38, 60, 61, 96, 163

Offshore Compliant Platforms: Analysis, Design, and Experimental Studies,
First Edition. Srinivasan Chandrasekaran and R. Nagavinothini.
© 2020 John Wiley & Sons Ltd.
This Work is a co-publication between John Wiley & Sons Ltd and ASME Press.

continuous ice crushing
  effect of ice parameters   140–145
  ice- vs. wave-induced responses
    145–147
  Korzhavin equation   135, 138
  response to ice loads   138–140
  spectrum   136–138
conventional TLP, 8
Coulomb friction   20–21
coupling effect   108
creep, ice failure modes   125
crushing, ice failure modes   125
current generation
  buoyant leg response   120–121
  deck response   116–119
  environmental loads   98–101
  tether tension variation   122

d

dampers   24
  metallic yield   27, 28
deck response
  to hydrocarbon fires   156–158
  to sea conditions   139
  of wind, waves, and current   116–119
  degrees of freedom   8, 99, 139, 167
  dynamic coupling motion   162
  dynamic ice force model   128
  dynamic vibration absorbers (DVAs)   35

e

elasticity   7–9, 132
elliptical buoyant legs
  conceptual development   179–182
  to wave action   182–184
Emil Simiu's wind velocity spectrum   15
energy-balance technique   10
environmental loads
  current generation   100–101
  random waves   97–98
  regular waves   96–97
  wind   98–100
explicit analysis solver   147

f

fast Fourier transform technique   17
fatigue damage   103, 108, 147, 151–153, 160, 173
finite element analysis (FEM)   63
floating liquefied natural gas (FLNG)   63
floating offshore wind turbines (FOWTs)   162, 163
floating platforms   2, 13, 36, 162
floating production storage and offloading (FPSO)   2
floating production system (FPS)   21, 63
floating production units (FPUs)   2
floating storage and offloading (FSO)   2
floating structures   16, 21–24, 162, 165
fourth-order Runge–Kutta differential equation   12
FPSO see floating production, storage, and offloading (FPSO)
friction dampers   26, 27
Froude scaling   38, 62, 64, 173, 174

g

glaciers   126
growlers   126
Gumbel method   24
guyed towers, AT, 19–21

h

hard tank   22
Harris spectrum   30
horizontal axis wind turbine (HAWT)   164
hydrocarbon fires   131–134
  deck response to   156–158
hydrodynamic buoyant mass damper (HBMD)   37
hydrodynamic forces   11, 14, 15

i

icebergs   126
ice cakes   126

Index

ice crushing strength   136, 138, 143, 145, 160
ice failure modes   127
ice fields   126
ice floes   126
ice force   127–129, 134, 135, 137–140, 144
ice force–time   135, 138
ice loads   126–129
  response to   138–140
ice sea conditions   138
  deck response to   139
  normal and extreme   135, 139, 147, 160
  PSD plots for   141–143
ice sheet   126–128, 134, 135 *see also* ice floe
ice speeds   128, 136
ice thickness   126, 127, 134–136, 138, 140–144, 160
ice velocity   128, 134–136, 138, 144
impact loads
  due to ship platform collisions   129–131
  impact response in Arctic region   154–155
  parametric studies   151–154
impact load–time history   149, 150
indenter shape   131, 153–154
indenter size   151–152
inverse fast Fourier transform (IFFT)   98

k

Kanai-Tajimi (K-T) power spectrum   18
Keulegan–Carpenter number   14, 15
knife-edge indenter   153
Korzhavin equation   135, 138

l

Lagrange equation   20, 31
Lagrangian coordinates   17, 21
linearization technique   17, 36
linear wave theory   17, 73

m

Mathieu equation   23, 85–87
Mathieu instability   10, 23
MATLAB, 15, 17, 23
mean ice force   136
mean sea level (MSL)   98, 147
meshing   147–148, 182
metallic yield dampers   27, 28
Miner's hypothesis   90
MLAT *see* multi-legged articulated tower (MLAT)
Morison equation   11–13, 15–19, 21–24, 36, 62, 95
multi-legged articulated tower (MLAT)
  with a TMD, 56–58
  without a TMD, 53–56
multiple degrees of freedom (MDOF) system   32
multi-tuned mass dampers (MTMDs)   34

n

Newmark-beta method   13, 14, 17, 18, 21, 23

o

offset condition, of TLP, 9
offshore triceratops   18

p

pack ice   126
pall friction damper   27
pancake ice   126
passive control algorithm   26–27
Pierson–Moskowitz (PM) spectrum   24, 145
piezoelectric accelerometers   65
pontoon (barge) type   165
postulated failure cases   85, 88–93
power spectral density (PSD)   15, 35, 77–82, 111, 113, 140, 167

q

quasi-static theory 12

r

radial and circumferential cracking, ice failure modes 125
rainflow-counting method 90
random sea conditions 97
random waves
  environmental loads 97–98
  response to 108–113
recentering 7
regular waves
  environmental loads 96–97
  response to 104–108
response amplitude operator (RAO) 14, 62, 97, 166
response control
  active control algorithm 25
  articulated tower (AT) of 48–53
  friction dampers 27
  metallic yield dampers 27
  passive control algorithm 26–27
  semi-active control algorithm 25–26
  tuned liquid column damper (TLCD) 30–31
  tuned liquid dampers 29–30
  viscous fluid dampers 27–28
return on investment (ROI)
ride-up effect 73
rotor nacelle assembly (RNA) 165

s

scaling law 173, 174
semi-active control algorithm 25–26
semi-compliant offshore structures 44
semi-submersible type 166
service life calculation 103
servo-hydraulic system 64
setdown 9
shallow-water wave theory 29, 30
shallow-water wind turbines 163
shear, ice failure 125
ship–platform collisions 160
  impact loads due to 129–131
shore-fast ice 126
single degree of freedom (SDOF) system 11, 32, 33
single-leg multi-hinged AT, 48
socket joint 18, 21
soft tank 22
spalling, ice failure modes 125
spar platforms 60
spar type 165
spring-mass system 37
STATELP, 15
stem bar 147
stiffened triceratops
  preliminary design 173–175
  response to wave action 175–177
  wave direction effect 177–179
Stokes wave theory 16
strength-based design approach 2
stress histogram 102
stringers, number of 154

t

taut-moored tethers 8, 60–61, 72, 87, 182
TBT *see* tethered buoyant tower (TBT)
tension leg platform (TLP) 8–19, 162
  BLSRP, 60
  construction worldwide 20
  mechanics 10
  response control of 38–44
  type 165
tethered buoyant tower (TBT) 62
tether fatigue analysis 101–103
tether response 140
tethers 165
  tension variation 122
thermal load 158

three-dimensional finite element analysis (3D FEA)   11
time-domain analysis   182
time-domain based potential theory   11
TLP *see* tension leg platform (TLP)
total force–time   184, 185
total ice force   127
transitional wind turbines   163
triceratops
   continuous ice crushing
      effect of ice parameters   140–145
      ice-*vs.* wave-induced responses   145–147
      Korzhavin equation   135, 138
      response to ice loads   138–140
      spectrum   136–138
   hydrocarbon fires   131–134
      deck response to   156–158
   ice crushing strength   143
   ice loads   126–129
      response to   138–140
   ice thickness   140–144
   ice velocity   144
   mass properties of   176
   response amplitude operator (RAO) plots   166
   response to impact loads   147–155
   rough sea conditions   184
   ship–platform collisions   129–131
   stiffened
      preliminary design   173–175
      response to wave action   175–177
      wave direction effect   177–179
   wind turbines   164, 166, 167
      free-decay response   166–169
      operable and parked conditions   169–170
      PSD plots   171–172
      tether response and service life estimation   172–173
      wave heading angles   170–171

TRSPAR, 23
tuned liquid column damper (TLCD)   27, 30–31
tuned liquid dampers (TLDs)   29–30, 36
   circular and rectangular   29
   MLAT with   56–58
   MLAT without   53–56
tuned mass dampers (TMDs)   27, 31–36

u

underwater tuned liquid column damper system (UWTLCD)   37
universal joint   19, 21, 44–46, 48, 53, 54

v

vertical axis wind turbine (VAWT)   164
viscous fluid dampers   27–28

w

wave heading angle   69, 170–171
waves   95
   buoyant leg response   120–121
   deck response   116–119
   environmental loads   98–100
   tether tension variation   122
Weibull distribution   13–14, 24
wind
   buoyant leg response   120–121
   deck response   116–119
   environmental loads   98–100
   tether tension variation   122
WindFloat   163
wind loads   62
wind power   163
wind turbines   161–162
   elliptical buoyant legs
      conceptual development   179–182
      to wave action   182–184
   evolution of   163–164
   stiffened triceratops

wind turbines (*cont'd*)
   preliminary design 173–175
   response to wave action 175–177
   wave direction effect 177–179
  support systems for
   pontoon (barge) type 165
   semi-submersible type 166
   spar type 165
   TLP type 165

triceratops 164, 166
  free-decay response 166–169
  operable and parked conditions 169–170
  PSD plots 171–172
  tether response and service life estimation 172–173
  wave heading angles 170–171